BEEKEEPING AND BEE CONSERVATION

BEEKEEPING AND BEE CONSERVATION

Patricio N. Magliano
Eric Jellen

Kruger Brentt
Publishers

2025

Kruger Brentt Publishers UK. LTD.
Company Number 9728962

Regd. Office: 68 St Margarets Road, Edgware, Middlesex HA8 9UU

ISBN: 978-1-78715-282-3

For information on all our publications visit our website at http://krugerbrentt.com/

PREFACE

"Beekeeping and Bee Conservation" is a comprehensive exploration of the vital relationship between humans and bees, offering insights into the practice of beekeeping and the conservation of these essential pollinators. This book serves as a valuable resource for beekeepers, conservationists, policymakers, and anyone interested in safeguarding the health and well-being of bees and their ecosystems.

Bees play a crucial role in pollinating a wide variety of crops, contributing to global food security, biodiversity, and ecosystem stability. However, bees face numerous threats, including habitat loss, pesticide exposure, disease, and climate change, which have led to declines in bee populations worldwide. In response, there is a growing need for effective strategies to support bee conservation efforts and promote sustainable beekeeping practices.

As editors of this volume, we have endeavored to compile a diverse array of perspectives, methodologies, and case studies from leading experts and practitioners in the fields of beekeeping and bee conservation. Through a blend of practical advice, scientific insights, and inspirational stories, this book covers a wide range of topics, including bee biology, hive management, pollinator-friendly gardening, habitat restoration, and community engagement.

Our overarching goal in presenting this book is to raise awareness about the importance of bees and inspire action to protect and conserve these invaluable pollinators. By highlighting the symbiotic relationship between humans and bees and providing practical guidance for sustainable beekeeping and conservation practices, we aim to empower individuals and communities to become stewards of bee health and biodiversity.

We extend our sincere gratitude to all the contributors who have generously shared their expertise, experiences, and insights in the fields of beekeeping and bee conservation. It is our hope that this book will serve as a catalyst for collaboration, innovation, and collective action in support of bees and their vital contributions to our world.

Patricio N. Magliano
Eric Jellen

CONTENTS

01 AN IN-DEPTH ANALYSIS OF THE HONEYBEE POPULATION IN SIBERIA, RUSSIA

1. INTRODUCTION: AN OVERVIEW

The honeybee was brought to Siberia approximately 230 years ago. The forest bee Apis mellifera mellifera L., also known as the Middle Russian race, was specifically bred and grown in Siberia due to its superior adaptation to the region's tough climatic circumstances. In recent decades, there has been a significant influx of bees from southern regions, including the Carpathian race (Apis mellifera carpatica) and the Caucasian gray mountain race (Apis mellifera caucasica Gorb.), into Siberia. This process has become pervasive and nearly unmanageable, resulting in a significant degree of interbreeding among bees. Currently, a significant issue in beekeeping throughout several countries is the extensive interbreeding of bees, resulting in a decrease in the diversity of native subspecies, the emergence of hybrids, and a decline in the genetic makeup of honeybees. Hybrid populations exhibit less adaptability to environmental settings that undergo rapid seasonal changes, resulting in increased susceptibility to illness and weakened immune responses [1–3].

Introgressive hybridization alters the genetic composition of indigenous honeybee populations, resulting in the erosion of their genetic distinctiveness [4]. The hybridization of distinct subspecies of honeybees can result in the disruption of established gene complexes, leading to a reduction in the adaptive characteristics of animals and populations, as well as changes in the biological and economically important indicators of bees. The extensive occurrence of hybridization among honeybees and the subsequent emergence of hybrid bees can undeniably facilitate the dissemination of diseases. The investigation of the scope of hybridization, the distinguishing features of hybrid bees, the examination of genetic mechanisms involved in hybridization, and the assessment of the impacts resulting from hybridization are subjects of significant interest. This study aims to characterize honeybees in Siberia by morphometric and molecular genetic analysis, specifically using mtDNA and microsatellite analysis. Additionally, the study aims to examine the infestation of bee colonies by Nosema.

2. MATERIALS AND METHODS

2.1. Region

Study was conducted to examine bees and bee colonies in three specific regions of Siberia: the Tomsk region, the Krasnoyarsk Krai, and the Altai Krai (as shown in Figure 1). The Tomsk region is situated in the geographical heart of Siberia, specifically in the southeastern portion of the West Siberian Plain. The meridian spans approximately 600 kilometers from its northern to southern bounds, resulting in a distinct difference in climate between the two locations. An inherent climatic feature of the northern region is a winter season that is both more intense and extended. The majority of the region's land is located inside the taiga zone, with woods occupying approximately 60% of the area. The climate is characterized as temperate continental, with significant daily and annual temperature variations, as well as prolonged winters lasting for around 5 to 6 months. The mean annual temperature is -0.6 °C, with July seeing an average temperature of +18.1 °C and January having an average temperature of 19.2 °C. The duration of the frost-free period ranges from 100 to 105 days. The amount of precipitation is 435 millimeters.

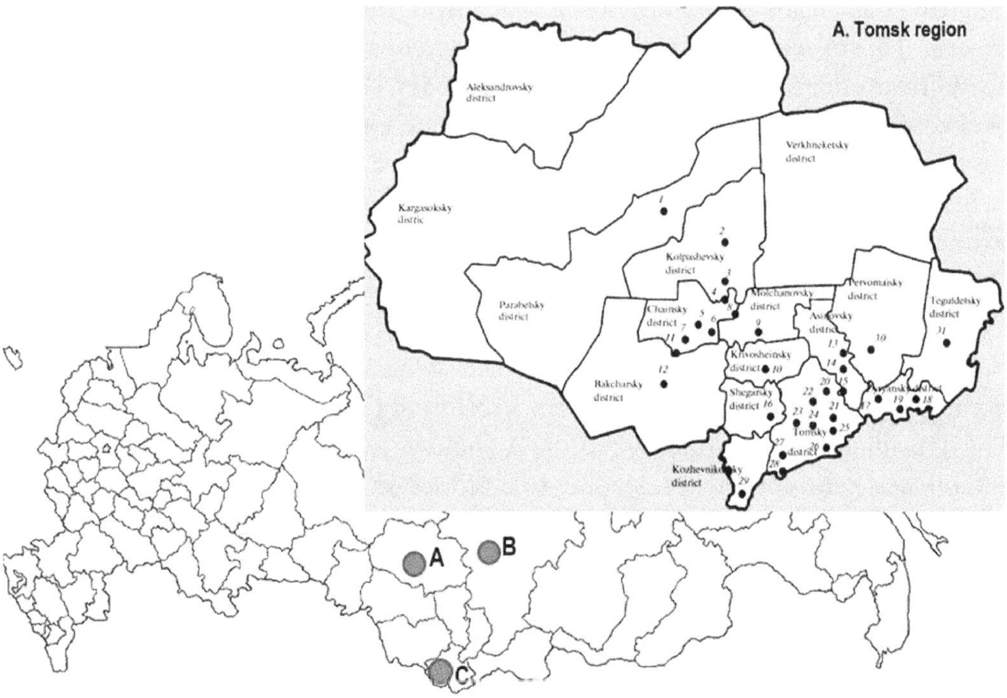

Figure 1. Map of localization of studied areas of Siberia (dots A–C) and apiaries of the Tomsk region (dots 1–31): A, the Tomsk region; B, the Krasnoyarsk Krai; C, the Altai Krai. 1, s. Parabel; 2, vicinity of g. Kolpashevo; 3, d. Novoabramkino; 4, s. Leboter; 5, s. Podgornoe; 6, d. Strelnikovo; 7, s. Gorelovka; 8, d. Sarafanovka; 9, s. Sokolovka, s. Mogochino; 10, s. Krivosheino; 11, s. Vysoky Yar, d. Krylovka; 12, s. Bakchar, s. Parbig; 13, d. Tihomirovka; 14, ur. Kuzherbak; 15, s. Novikovka; 16, s.

Kargala; 17, s. Dubrovka; 18, s. Okuneevo; 19, s. Zyryanskoe; 20, d. Kuskovo; 21, p. Zarechnyi (Mezheninovskoe rural settlement); 22, d. Bodazhkovo, s. Semiluzhki, p. Zarechnyi (Malinovskoe rural settlement); 23, d. Nizhne-Sechenovo, d. Berezkino, s. Zorkaltsevo, s. Rybalovo, d. Kudrinsky uchastok, d. Gubino; 24, p. Sinii Utes, d. Magadaevo, d. Prosekino, s. Kolarovo, vicinity of Tomsk; 25, d. Bolshoe Protopopovo; 26, s. Mezheninovka; 27, d. Kandinka, s. Kurlek; 28, s.Yar; 29, d. Elovka; 30, d. Krutolozhnoe; 31, s. Teguldet. Apiaries located at a distance less than 15 km from each other are marked as a single point.

Krasnoyarsk Krai is situated in Eastern Siberia. The climate is predominantly continental, with forests covering 70% of the country. The Altai Krai is situated in the southeastern region of Western Siberia. The region encompasses nearly all of Russia's diverse ecological zones, including the steppe, forest steppe, taiga, and mountains. The climate in the Altai Territory is characterized by significant heterogeneity due to a variety of geographical factors. The foothills experience a temperate climate that gradually transitions into a continental climate.

2.2. Samples

The samples are collected from various geographical regions within the Tomsk region, which exhibit ecological and climatic diversity. These regions include areas with significant beekeeping activity (southern districts) as well as areas with minimal beekeeping activity (northern districts), as determined by the expertise of specialists from the Society of Beekeepers. The honeybees from the apiaries located in the Krasnoyarsk Krai and the Altai Krai were also examined for the purpose of comparison. The study examined a total of 332 bee colonies (consisting of 60 apiaries) from Siberia. The investigation involved the use of morphometric analysis on 3043 honey bee workers, as well as molecular genetic methods on 2073 bees through mtDNA analysis and microsatellite analysis on a range of 252 to 515 bees. Specifically, the study included 318 bee colonies from the Tomsk region, 10 colonies from the Krasnoyarsk Krai, and 5 colonies from the Altai Krai (Figure 1).

The bee colonies from the Krasnoyarsk Krai were obtained from a distinct isolated population of Old Believers, which had been present in the forest for over 60 years without introducing new honeybees. Bee colonies originating from the Altai Krai region have been gathered and housed at an apiary situated in the foothills. A study was conducted on the infestation of bee colonies by Nosema infections. A total of 1983 samples were gathered from 132 bee colonies in 68 apiaries in Siberia between 2012 and 2015.

2.3. Morphometric method

The study focused on analyzing morphometric characteristics related to wing venation, specifically the cubital index, the hantel index, and the discoidal shift. These factors were examined as shown in Figure 2.

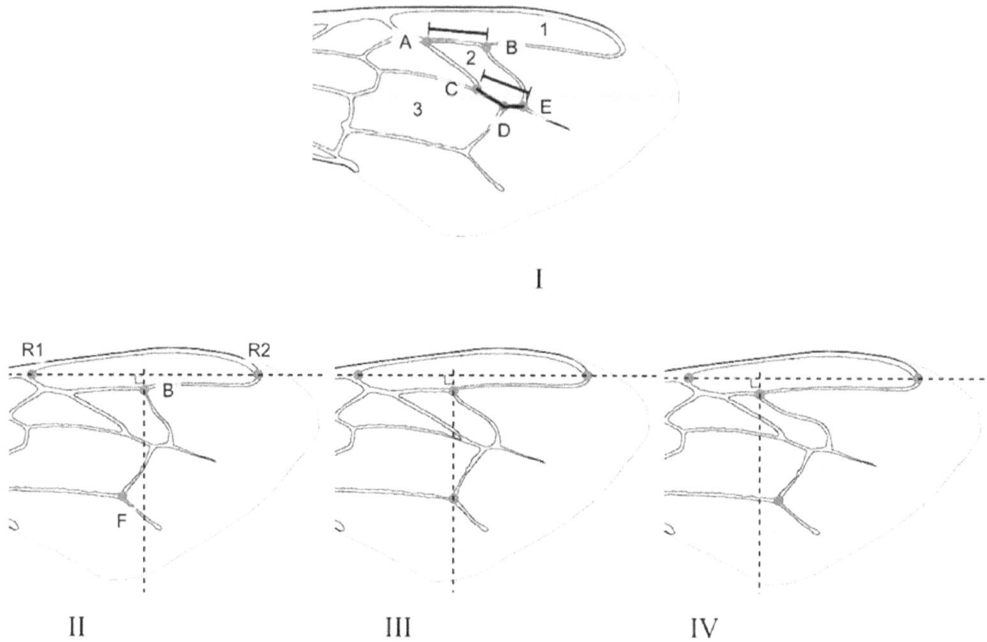

Figure 2. Scheme of the front wing venation of honeybee (I) and discoidal shift (II, III, and IV), showing the position of the horizontal and vertical lines (dashed lines). A, B, C, D, and E—the key points and segments that are used in determining the wing index (cubital index: CD/DE; hantel index: CE/AB). Options of discoidal shift: II—negative (point F is located to the left of the perpendicular line); III—zero (point F located on a perpendicular line); IV—positive (point F is located to the right of the perpendicular line). Designation of sells: 1, radial; 2, cubital; 3, discoidal.

2.4. Analysis of mitochondrial DNA (mtDNA)

DNA isolation and polymerase chain reaction (PCR) was carried out according to standard techniques with some modifications [5,6]. To amplify the COI–COII mtDNA locus, the following sequences of primers were used: 3′-CACATTTAGAAATTCCATTA, 5′-ATAAATATGAATCATGTGGA [5]. Amplification products were fractionated in 1.5% agarose gel, and the results were documented with the use of Gel-Doc XR+.

2.5. Microsatellite analysis

Microsatellite analysis refers to the process of studying and analyzing microsatellites, which are short, repetitive DNA sequences found in the genome of an organism. Variability of eight microsatellite loci was studied: A008 (=A8), Ap049, AC117, AC216, Ap243, H110, A024, and A113. PCR was performed using specific primers and reaction conditions according to Solignac et al. [7]. Amplification products were analyzed with ABI Prism 3730 Genetic Analyser (Applied Biosystems, Inc., Foster City, CA) and GeneMapper Software (Applied Biosystems, Inc.). Two microliters of PCR products were mixed with GeneScan500-ROX size standards

(Applied Biosystems, Inc.) and deionized formamide. Samples were run according to the manufacturer's recommendations. These genetic parameters were calculated: allelic frequencies and standard error.

2.6. Infestation of honeybees by *Nosema*

A random sample of bees ranging from 10 to 70 was collected from each bee colony and subjected to examination for the presence of Nosema. Prior to testing, bee samples were kept at room temperature in 70% (v/v) ethanol. An individual analysis was conducted for each bee. The midgut of each sample was dissected, with one portion being examined for the presence of Nosema spores using a light microscope, and the other portion being utilized for DNA extraction. The midgut was immersed in 200 µL of distilled water and inspected using dark-field microscopy to detect the presence of Nosema spores [8]. The DNA from the midgut was isolated using a DNA purification kit called PureLink™ Mini (Invitrogen, Carlsbad, CA), following the instructions provided by the manufacturer.

After extraction, the samples were submitted to duplex-PCR [9,10]. The primer sequences utilized to amplify the 218-bp fragment corresponding to the 16S ribosomal gene of *N. ceranae* were 218MITOC–FOR 5'–CGGCGACGATGTGATATGAAAATATTAA–3' and 218MITOC–REV 5'–CCCGGTCATTCTCAAACAAAAAACCG–3'[9]. The primer sequences used to amplify the 321 bp fragment corresponding to the 16S ribosomal gene of *N. apis* were 321APIS–FOR 5'–GGGGGCATGTCTTTGACGTACTATGTA–3' and 321APIS–REV 5'–GGGGGGCGTTTAAAATGTGAAACAACTATG–3'[9]. PCR was performed using specific primers and reaction conditions according to Hamiduzzaman et al. [10]. PCR products were analyzed on 1.5 % (m/v) agarose gels and visualized using UV illumination (Gel Doc XR+, BioRad, Foster City, CA, USA). All analyses were carried out in duplicate, positive and negative controls were used, and identical results were obtained.

Furthermore, DNA sequencing was employed to confirm the presence of selected pieces from both N. ceranae and N. apis, in addition to the utilization of particular primers and fragment size for species identification. The sequencing process was conducted in both the forward and reverse directions using either the forward or reverse primer. The BigDye Terminator v3.1 Cycle Sequencing Kit from Applied Biosystems in Foster City, CA was used for this purpose. The DNA sequencing was conducted using the ABI Genetic Analyzer 3730 from Applied Biosystems, following the procedure provided by the vendor.

3. RESULTS AND DISCUSSION

We conducted a molecular genetic examination of bee colonies using mtDNA analysis (specifically, the COI-COII locus). We collected 5-6 samples from each bee colony to determine the maternal lineage and origin of the bee colony.

3.1. Genetic diversity of COI-COII mtDNA locus

A study was done to evaluate the genetic variation of the COI-COII mtDNA region in honeybee populations from the Tomsk region. More information may be found in reference [11]. Three different forms of the COI-COII mtDNA locus were documented: PQQ, PQQQ (characteristic of the Middle Russian race), and Q (characteristic of southern races). Our findings indicate that 64% of bee colonies on the maternal line are derived from the Middle Russian race, 28% originate from southern subspecies, and 8% are comprised of mixed bee colonies. In the Tomsk region, the southern areas, where beekeeping is common, exhibit greater genetic variety in honeybees compared to the northern regions. In the northern districts, the majority of bee colonies (96%) and apiaries (73%) have a uniform genetic variant at the COI-COII locus. The bee colonies originating from the Middle Russian breed exhibited genetic heterogeneity at the COI-COII locus. The PQQ variant was found in 86.1% of the total number of bee colonies of the Middle Russian race, while PQQQ was found in 9.4%. Additionally, 4.5% of bee colonies contained individuals with both the PQQ and PQQQ alleles.

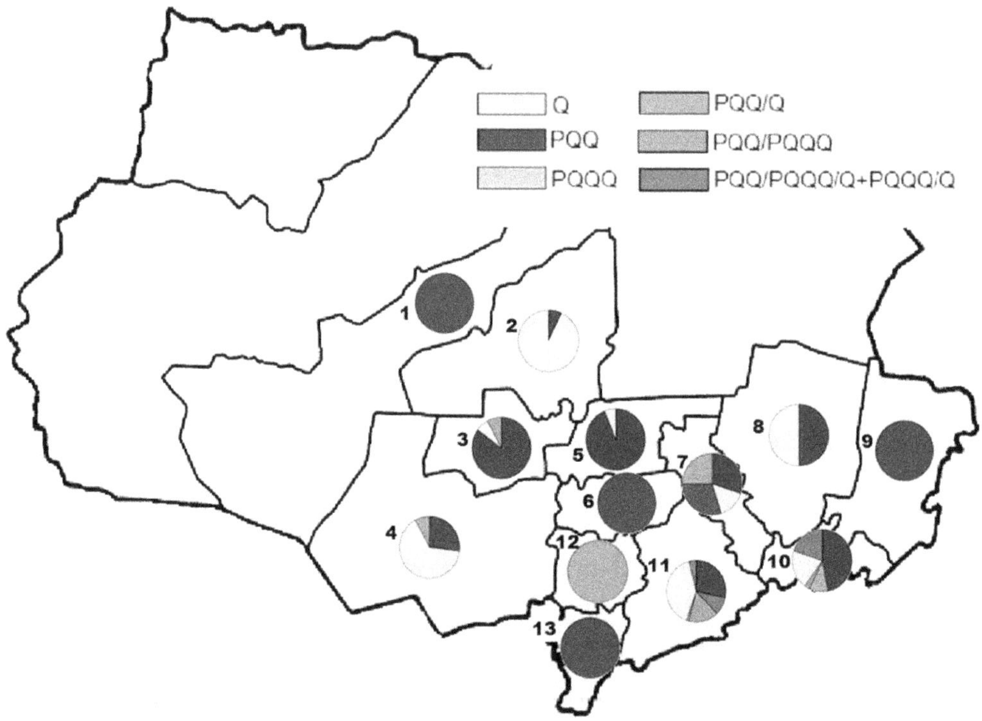

Figure 3. Distribution of COI-COII mtDNA locus variants for the districts (numbers 1–13) of the Tomsk region. Northern districts: 1, Parabelsky; 2, Kolpashevsky; 3, Chainsky; 4, Bakcharsky; 5, Molchanovsky; 6, Krivosheinsky; and southern districts: 7, Asinovsky; 8, Pervomaisky; 9, Teguldetsky; 10, Zyryansky; 11, Tomsky; 12, Shegarsky; 13, Kozhevnikovsky. Variants PQQ/PQQQ/Q (1%) and PQQQ/Q (3%), which are found only in the Tomsk district, are combined.

Based on the analysis of mtDNA (locus COI-COII), assessment of the genetic diversity of the honeybee in apiaries of the Tomsk region has shown that the genetic structure of bee populations in the Tomsk region is complex and mosaic, especially in the southern parts of the region (Figure 3). No large areas with an array of bees having a homogeneous genetic (race) composition and maternally originating from the Middle Russian race have been found; a few apiaries were revealed, in which all bees originated from the Middle Russian breed.

In the study of variability of the COI-COII mtDNA locus in honeybees from apiaries of the Krasnoyarsk Krai and the Altai Krai, two variants of the COI-COII locus specific for Middle Russian race were identified: only variant PQQ was registered in honeybees of Krasnoyarsk Krai (Yenisei population) and two variants (PQQ and PQQQ) were found in honeybees from the Altai Krai. No a variant Q specific for southern races of bee was detected.

Due to the fact that mtDNA analysis allows assessing only the maternal component in the genome of the honeybee, bee colonies were investigated by the morphometric analysis to identify the characteristics of both the maternal and paternal lines, and to assess the level of hybridization.

3.2. Morphometric study of honeybees

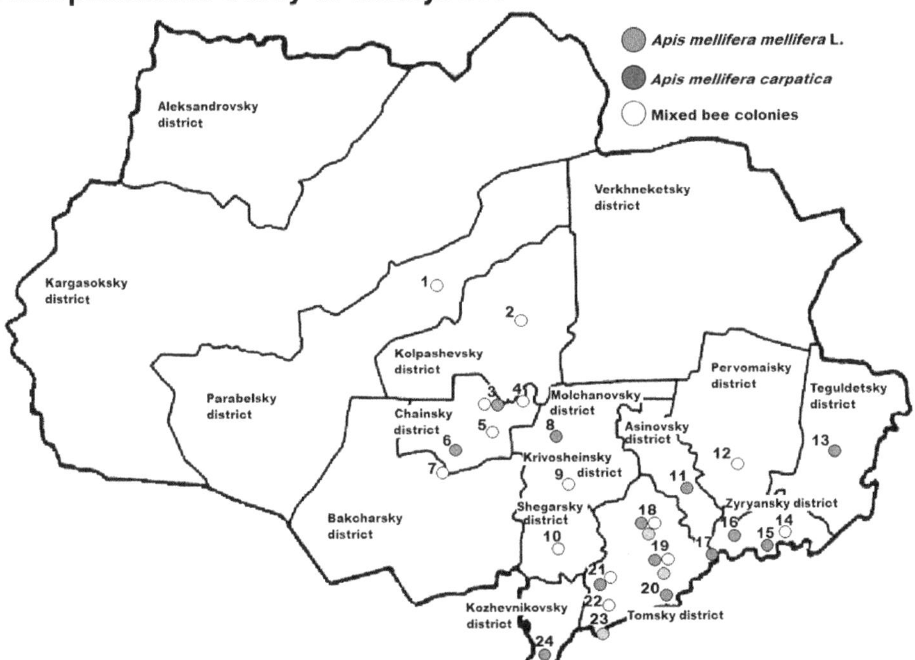

Figure 4. Distribution of subspecies and hybrids in the apiaries of the Tomsk region on the basis of cubital index of bee workers. Studied settlement are indicated by numbers: 1, s. Parabel; 2, vicinity of g. Kolpashevo; 3, Podgornoe; 4, s. Leboter; 5, d. Strelnikovo; 6, s. Gorelovka; 7, s. Vysoky Yar, d. Krylovka; 8, s.

Mogochino; 9, s. Krivosheino, Sokolovka; 10, s. Kargala; 11, ur. Kuzherbak; 12, d. Krutolozhnoe; 13, s. Teguldet; 14, s. Okuneevo; 15, s. Zyryanskoe; 16, s. Dubrovka; 17, s. Novorozhdestvenskoe; 18, s. Kornilovo, s. Semiluzhki, p. Zarechnyi (Malinovskoe rural settlement); 19, p. Sinii Utes, d. Magadaevo, d.6 Prosekino, s. Kolarovo, vicinity of Tomsk; p. Zarechnyi (Mezheninovskoe rural settlement); 20, s. Mezheninovka; d. Arkashovo; 21, s. Zorkaltsevo, s. Rybalovo, d. Kudrinsky uchastok, d. Gubino; 22, s. Kurlek; 23, s.Yar; 24, d. Elovka. Apiaries located at a distance less than 15 km from each other are marked as a single point.

The findings of the morphometric analysis of honeybees from the investigated areas of Siberia (namely, the Tomsk region, the Krasnoyarsk Krai, and the Altai Krai) exhibited variations. Based on the morphometric analysis, most of the bee colonies in the Tomsk region are a combination of the Middle Russian race of bees and bees from the southern region, primarily the Carpathian race. Figure 4 displays data regarding the distribution of subspecies and hybrids in the apiaries in the Tomsk region, based on the cubital index. Several apiaries that breed Middle Russian bees were discovered in both the northern and southern regions of the Tomsk area.

The bee colonies acquired from remote apiaries in the Krasnoyarsk Krai region are of significant importance. The region containing these remote apiaries remained unaffected by other honeybee subspecies for an extended period, and all the examined bees possessed solely the PQQ variation of the COI-COII mtDNA locus. Nevertheless, upon comparing the morphometric data of bees from isolated apiaries with the established Russian and European standards for Apis m. mellifera, it was observed that the lower limit values of the cubital index were reduced in the studied bees. Consequently, a majority of bee colonies exhibited a deviation from the average values of the cubital index, as indicated in Table 1. There are multiple potential justifications for the outcomes. Firstly, this phenomenon could be attributed to genetic drift, which is caused by the isolation of these apiaries and the limited population of bees. The significant range of variation in the cubital index is attributed to the adaptation to more extreme climatic conditions in the environment. However, these remote apiaries in the Krasnoyarsk Krai can be regarded as a distinct population of the Middle Russian bee that has existed for a significant period without having any impact on other subspecies of honeybee.

The morphometric study results revealed that the bee colonies of the Altai population originated from the Middle Russian race, but there was some evidence of influence from the southern races. As an illustration, the parameter "Discoidal shift" diverges from the established criteria for the Russian breed: bee colony No. 7 exhibited individuals with both positive and zero values for discoidal shift, as indicated in Table 2. Bee colonies from the Krasnoyarsk Krai are sourced from remote, sparsely populated areas in the taiga. In contrast, bee colonies from the Altai Krai are found in an area with a well-developed beekeeping industry and a continuous influx of bees from various sources.

Table 1. Morphometric parameters (wing venation) of honeybee workers from 10 bee colonies of the Krasnoyarsk Krai (Yenisei population).

Geographical location: settlement	Bee colony, №	Cubital index, standard units		Hantel index, standard units		Discoidal shift, %		
		Lim: min max	*M ± m*	*Lim: min max*	*M ± m*	–	0	+
Ostyatskoe	1	1.24 2.00	1.61±0.04	0.675 0.892	0.795±0.011	100.0	0	0
	2	1.39 1.74	1.51±0.02	0.743 0.912	0.849±0.012	83.3	16.7	
	3	1.23 1.74	1.51±0.03	0.736 0.883	0.837±0.008	83.3	16.7	0
	4	1.20 1.67	1.45±0.02	0.723 0.900	0.837±0.009	97.0	3.0	0
	5	1.24 1.79	1.46±0.03	0.735 0.923	0.842±0.010	87.0	13.0	0
Kolmogorovo	1	1.32 2.10	1.60±0.05	0.724 0.900	0.820±0.009	97.0	3.0	0
	2	1.12 1.76	1.51±0.03	0.758 0.919	0.845±0.008	93.0	7.0	0
	3	1.28 1.86	1.56±0.04	0.746 0.985	0.810±0.011	97.0	3.0	0
	4	1.07 1.76	1.45±0.04	0.716 0.945	0.830±0.011	97.0	3.0	0
Yaksha	1	1.31 1.85	1.59±0.02	0.711 0.846	0.775±0.008	100.0	0	0
Standard for *Apis mellifera mellifera*								
I		1.30 2.10	1.70	0.600 0.923	No data	No data		
II		1.30 1.90	1.5 to 1.7	0.600 0.923		91– 100	5–10	0.00

Minimum 30 samples from each bee colony were studied.

Lim, limits of value of the sing; *M* ± *m*, average value of the sign ± the standard error of the mean; I, European breed standard based on values of cubital and hantel indexes [12]; II, Russian breed standard.

3.3. The correlation between morphometric parameters and mtDNA analysis data in honeybees in Siberia.

The diagnostic results of honeybees' external morphological features, including the cubital index, hantel index, and discoidal shift, were obtained from 11 bee colonies with different variants of the COI-COII mtDNA locus. These results are provided in Table 2. Out of the 11 bee colonies, only 4 demonstrated complete adherence to the breeding criterion based on morphometric and mtDNA analysis. These four colonies include three colonies of Apis mellifera mellifera and one family of Apis mellifera carpatica. Out of the total number of colonies, only seven are hybrid. Among these seven colonies, three showed a notable disparity in genetic and morphometric parameters. Therefore, while attempting to identify the breeds in the context of widespread bee hybridization, it is crucial to take into account not only the characteristics of mtDNA, but also morphometric data, with the discoidal shift likely being the most significant.

This data aligns with the findings of the hybrid apiary study, which has been selectively breeding the Middle Russian bee for over 30 years. However, in the past decade, there has been a significant influx of southern bee races [6]. Over 50% of individuals are identified as belonging to the southern races based on mtDNA study using variant Q of the locus COI-COII, also known as the "southern" mitotype. However, none of these individuals matched the characteristics of the southern race as determined by morphometric study (Table 3). 33% of individuals with a "southern" mitotype exhibited two distinct morphometric traits that are specific to the Middle Russian race. Approximately 6% of the bees belonging to the Middle Russian race (variant PQQ of the locus COI-COII) exhibited complete agreement between their mitotype and morphometric parameters. 18% of the bees exhibited a mitotype and two morphometric parameters that were unique to the Middle Russian bees. This shows a pattern of hybridization between Middle Russian and southern breeds on this apiary. However, the process of gene displacement varies for bees of different origins. For bees of the Middle Russian race, the scale of gene displacement is smaller. This is because a smaller percentage of bees with "southern" morphometric characteristics were observed among individuals with variant PQQ, compared to bees with "southern" mitotypes.

Table 2. Morphometric parameters (wing venation) of honeybee workers of 11 bee colonies from apiaries of Siberia.

region	Geographical location District	Settlement	Bee colony, №	Number of studied bees	Sequence composition of the COI-COII mtDNA locus	Cubital index, standard units *Lim: min max*	*M*	*sd*	Hantel index, standard units *Lim: min max*	*M*	*sd*
	Apis mellifera mellifera[*]										
Tomsk region	Tomsky	p. Zarechnyi	1	30	PQQQ	1.39 2.23	1.66	0.216	0.712 0.932	0.826	0.052
		s. Kurlek	2	28	PQQQ	1.74 3.29	2.14	0.376	0.857 1.053	0.937	0.055
	Zyryansky	s. Dubrovka	3	30	PQQ	1.43 2.47	1.69	0.232	0.672 0.933	0.849	0.060
	Molchanovsky	s. Mogochino	4	30	PQQ	1.26 2.56	1.92	0.290	0.806 1.000	0.879	0.055
			5	43	PQQ	1.36 2.00	1.73	0.181	0.693 0.926	0.821	0.038
Altai Krai	Zmeinogorsky	Vicinity of c. Zmeinogorsk	6	29	PQQ	1.19 2.00	1.55	0.232	0.758 0.967	0.858	0.062
			7	30	PQQQ	1.50 2.50	1.80	0.245	0.722 0.984	0.845	0.059
Krasnoyarsk Krai	Yeniseisky	p. Yaksha	8	30	PQQ	1.31 1.85	1.59	0.132	0.711 0.846	0.775	0.044

region	Geographical location: District	Settlement	Bee colony, №	Number of studied bees	Sequence composition of the COI-COII mtDNA locus	Cubital index, standard units: *Lim: min max*	*M*	*sd*	Hantel index, standard units: *Lim: min max*	*M*	*sd*
	Southern breeds[*]										
Tomsk region	Tomsky	s. Semiluzhki	9	50	Q	1.68 3.64	2.51	0.374	0.867 1.210	1.050	0.047
		s. Kurlek	10	29	Q	1.30 2.29	1.66	0.220	0.735 0.965	0.878	0.060
		p. Sinii Utes	11	30	Q	1.83 2.87	2.37	0.334	0.815 1.053	0.931	0.065
Standart of breeds	*A. m. mellifera*[**]		PQQ, PQQQ and other			1.30 2.10	1.70	–	0.600 0.923	–	–
	A. m. mellifera[***]					1.30 1.90	1.6	–	0.600 0.923	–	–
	A. m. carnica[**]		Q			2.40 3.00	2.7	–	≥ 0.925	–	–
	A. m. caucasica[**]		Q			1.70 2.30	2.0	–	No data	–	–

Lim, limits of values; M, arithmetic mean; *sd*, standard deviation.

[*]Breed indicated according to the data of mtDNA analysis.

[**]European breed standard based on values of cubital and hantel indexes [12].

[***]Russian breed standard. Discoidal shift are given according to Russian standards.

Table 2. Continued

region	Geographical location District	Settlement	Bee colony, №	Number of studied bees	Sequence composition of the COI- COII mtDNA locus	Discoidal shift, % –	0	+
	Apis mellifera mellifera *							
Tomsk region	Tomsky	p. Zarechnyi	1	30	PQQQ	73.30	26.70	0.00
		s. Kurlek	2	28	PQQQ	32.10	53.60	10.70
	Zyryansky	s. Dubrovka	3	30	PQQ	73.33	26.67	0.00
	Molchanovsky	s. Mogochino	4	30	PQQ	70.00	30.00	0.00
			5	43	PQQ	100.0	0.00	0.00
Altai Krai	Zmeinogorsky	Vicinity of c. Zmeinogorsk	6	29	PQQ	94.00	6.00	0.00
			7	30	PQQQ	46.70	46.70	6.60
Krasnoyarsk Krai	Yeniseisky	p. Yaksha	8	30	PQQ	100.0	0.00	0.00
	Southern breeds*							
Tomsk region	Tomsky	s. Semiluzhki	9	50	Q	4.00	20.00	76.00
		s. Kurlek	10	29	Q	72.40	27.60	0.00
		p. Sinii Utes	11	30	Q	6.70	76.70	16.70
Standart of breeds	*A. m. mellifera* **				PQQ, PQQQ and other	–	–	–
	A. m. mellifera ***					91–100	5–10	0.00
	A. m. carnica **				Q	0–5	0–20	80–100
	A. m. caucasica **				Q	60–70	20–30	3–5

Lim, limits of values; M, arithmetic mean; *sd*, standard deviation.

*Breed indicated according to the data of mtDNA analysis.

**European breed standard based on values of cubital and hantel indexes [12].

***Russian breed standard. Discoidal shift are given according to Russian standards.

Table 3. The accordance of morphometric parameters in individuals with different genetic variants of the COI-COII mtDNA locus (see details in reference [6]).

mtDNA		Variant PQQ		Variant Q	
Number of studied bees, %		44.44		55.56	
Race		*Apis mellifera mellifera*	Southern race	*Apis mellifera mellifera*	Southern race
The combination of features characteristic for different races	3 parameters $x^1 + x^2 + x^3$	5.6	7.4	7.4	0.0
	2 parameters, total, including $x^1 + x^2$ $x^1 + x^3$ $x^2 + x^3$	18.5 1.9 3.7 13.0	13.0 1.9 11.1 0	33.3 1.9 0 31.5	14.8 11.1 3.7 0
	1 parameter, total	13.0	18.5	14.8	33.3
x^1, x^2, x^3 are parameters of cubital index, hantel index, and discoidal shift, respectively.					

Therefore, the findings from the investigation of hybrid apiaries and bee colonies demonstrate the significance and requirement for a thorough methodology in accurately identifying different honeybee breeds. However, the data have scientific significance for studying genetic processes during the hybridization of several subspecies of honeybees and for assessing the phenomenon of gene replacement, when genes from one race are replaced by genes from another race. An intriguing case of hybridization is observed between the Middle Russian bee and Carpathian bee, as these races belong to distinct evolutionary lineages. Microsatellite loci are the most informative molecular genetic markers for these types of investigations. Microsatellite markers have multiple applications in the analysis of genetic structure in honeybee populations and colonies. They are valuable for assessing genetic diversity, studying introgressive hybridization, distinguishing between different subspecies (ecotypes), establishing evolutionary relationships, identifying adaptive features within the four evolutionary branches (A, M, C, and O), mapping quantitative trait loci (QTL), and identifying genetic markers linked to economically important traits [3,7,13–46]. The initial stage of any research involving honeybee populations is the characterization of the allele spectrum of microsatellite loci and the examination of their variability in subspecies, colonies, and individuals.

3.4. Microsatellite analysis

The study focused on examining the variability of eight microsatellite loci (A008, Ap049, AC117, AC216, Ap243, H110, A024, and A113) in honeybees from the Siberian region. Out of the seven loci examined, only the AC216 locus showed a homozygous genotype in all of the bees investigated, with an allele size of 91 base pairs. The genotypes and alleles' range and frequency were established for each locus (Table 4).

Table 4. Characterization of variability of seven microsatellite loci in honeybees from Siberia.

Locus	Genotype	Frequency of genotype	Allelic frequency with an error
A008	152–152	0.006	P_{152}=0.0311±0.0054 P_{162}=0.8049±0.0123 P_{166}=0.0010±0.0031 P_{168}=0.0010±0.0031 P_{170}=0.0213±0.0045 P_{172}=0.0243±0.0048 P_{174}=0.0825±0.0086 P_{176}=0.0029±0.0017 P_{178}=0.0262±0.0050 P_{180}=0.0039±0.0019
	152–162	0.049	
	152–170	0.002	
	162–162	0.736	
	162–168	0.002	
	162–170	0.016	
	162–172	0.039	
	162–174	0.033	
	166–172	0.002	
	170–170	0.006	
	170–174	0.016	
	172–172	0.004	
	174–174	0.037	
	174–176	0.004	
	174–178	0.031	
	174–180	0.008	
	176–178	0.002	
	178–178	0.010	
	n=515		
Ap049	118–127	0.002	P_{118}=0.0010±0.0001 P_{121}=0.0069±0.0025 P_{127}=0.6581±0.0149 P_{130}=0.1759±0.0120 P_{139}=0.1403±0.0109 P_{142}=0.0010±0.0001 P_{152}=0.0168±0.0040
	121–127	0.002	
	121–130	0.006	
	121–139	0.006	
	127–127	0.529	
	127–130	0.187	
	127–139	0.053	
	127–152	0.019	
	130–130	0.055	
	130–139	0.045	
	130–152	0.002	
	139–139	0.081	
	139–152	0.013	
	142–152	0.002	
	152–152	0.002	
	n=506		
AC117	175–175	0.008	P_{175}=0.0910±0.0092 P_{179}=0.0879±0.0090 P_{183}=0.8211±0.0123
	175–179	0.020	
	175–183	0.145	
	179–179	0.012	
	179–183	0.131	
	183–183	0.683	

Locus	Genotype	Frequency of genotype	Allelic frequency with an error
	n=489		
H110	162–162	0.567	P_{162}=0.7522±0.0167 P_{166}=0.0627±0.0093 P_{170}=0.1851±0.0150
	162-166	0.116	
	162–170	0.254	
	166–166	0.003	
	166–170	0.003	
	170–170	0.057	
	n=335		
n, number of studied samples is indicated in bold.			

Locus	Genotype	Frequency of genotype	Allelic frequency with an error
Ap243	255–255	0.401	P_{255}=0.5278±0.0222 P_{263}=0.3175±0.0207 P_{269}=0.0833±0.0123 P_{272}=0.0635±0.0109 P_{275}=0.0079±0.0039
	255–263	0.167	
	255–269	0.056	
	255–272	0.028	
	255–275	0.004	
	263–263	0.175	
	263–269	0.075	
	263–272	0.040	
	263–275	0.004	
	269–269	0.004	
	269–272	0.028	
	272–272	0.012	
	272–275	0.008	
	n=252		
A024	94–94	0.344	P_{94}=0.4736±0.0186 P_{96}=0.1014±0.0112 P_{98}=0.0375±0.0070 P_{100}=0.0194±0.0051 P_{102}=0.2097±0.0152 P_{104}=0.1528±0.0134 P_{106}=0.0056±0.0028
	94–98	0.036	
	94–100	0.033	
	94–102	0.175	
	94–104	0.014	
	96–96	0.067	
	96–104	0.058	
	96–106	0.011	
	98–98	0.019	
	100–100	0.003	
	102–102	0.089	
	102–104	0.067	
	104–104	0.083	
	n=360		

Locus	Genotype	Frequency of genotype	Allelic frequency with an error
A113	208–212	0.003	P_{208}=0.0013±0.0013 P_{210}=0.0144±0.0043 P_{212}=0.2350±0.0153 P_{214}=0.0026±0.0018 P_{218}=0.5953±0.0177 P_{220}=0.1084±0.0112 P_{222}=0.0013±0.0013 P_{224}=0.0013±0.0013 P_{226}=0.0183±0.0048 P_{228}=0.0196±0.0050 P_{232}=0.0026±0.0018
	210–210	0.003	
	210–218	0.021	
	210–220	0.003	
	212–212	0.177	
	212–214	0.005	
	212–218	0.078	
	212–220	0.013	
	212–222	0.003	
	212–226	0.005	
	212–228	0.003	
	212–232	0.005	
	218–218	0.475	
	218–220	0.117	
	218–226	0.021	
	218–228	0.003	
	220–220	0.018	
	220–224	0.003	
	220–226	0.010	
	220–228	0.034	
	n=383		

The variability of microsatellite loci varied: loci AC117 and H110 had the lowest number of alleles (3 alleles), while loci A008 (10 alleles) and A113 (11 alleles) had the highest number of alleles. Simultaneously, all but one of the polymorphic loci (excluding locus A024) exhibited a dominant allele with a frequency exceeding 0.5. The frequencies ranged from 0.5278 for allele "255" of locus Ap243 to 0.8211 for allele "183" of locus AC117, regardless of the number of alleles found. In order to determine the characteristics of honeybees from various geographical regions, we conducted a comparative analysis of the variability of specific genetic markers in four populations of Apis mellifera mellifera (also known as the dark-colored forest bee or Middle Russian race) from Siberia, the Urals, and Europe. Our analysis included data from our own research in the Tomsk region and Krasnoyarsk Krai, as well as information from relevant literature sources [15,16,47] (refer to Tables 5 and 6). The Ural population, also known as the Bashkir population, found within the nature reserve, consists of a distinctive group of the dark-hued forest bee, specifically the Burzyan bee.

Table 5. Allele frequency at three loci in honeybees from different geographic areas of Russia and Europe.

Locus	Alleles (pb)	Allelic frequency					
		Russia			Europe**		
		Krasnoyarsk Krai	Tomsk region	Ural* (Bashkortostan)	Belgium (Chimay)	Sweden (Umea)	France (eight geographic areas)
A008	148				0.783	0.727	0.267–0.969
	152		0.006				
	154			0.897			0–0.083
	155						0–0.033
	156			0.053	0.133	0.227	0.017–0.300
	157						0–0.050
	158			0.053		0.023	0–0.117
	159						0–0.017
	160				0.050		0–0.100
	162	1.000	0.912		0.033		0–0.034
	164					0.023	0–0.020
	166		0.003				0–0.017
	170		0.003				
	172		0.032				
	174		0.044				
N		**120**	**170**	**48**	**60**	**44**	**634**
A024					No data	No data	
	94	0.216	0.741				
	96	0.358					
	98	0.132		0.896			0.804
	100	0.034	0.020				
	102	0.025	0.227				
	104	0.216	0.012				
	106	0.020		0.104			0.130
	108						0.065
N		**102**	**172**	**48**			**46**

Locus	Alleles (pb)	Allelic frequency					
		Russia			Europe**		
		Krasnoyarsk Krai	Tomsk region	Ural* (Bashkortostan)	Belgium (Chimay)	Sweden (Umea)	France (eight geographic areas)
A113	202				0.083	0.024	0.017–0.267
	204						
	208						0–0.017
	210	0.021	0.009				
	212		0.174				0–0.030
	214		0.006		0.033		0.010–0.500
	216			0.063			0–0.017
	218	0.898	0.540	0.865			0–0.020
	220	0.081	0.183	0.042	0.833	0.857	0.433–0.810
	222		0.003	0.032		0.024	0–0.041
	224		0.003		0.017	0.048	0–0.060
	226		0.040			0.048	0–0.034
	228		0.043		0.017		0.017–0.071
	230						0–0.052
	232						0–0.017
	234				0.017		0–0.017
	236						0–0.020
	238						0–0.017
	240						0–0.010
N		**118**	**175**	**48**	**60**	**44**	**634**

*data from reference [47].

**data from references [15,16].

N, number of studied samples.

The minimum and maximum values of allelic frequencies represented for loci A008 and A113 in honeybees of France populations; allelic frequencies for locus A024 are given for bees of only Northern France population.

Table 6. Allele frequency at three loci in honeybees from different populations of Russia.

Locus	Alleles (pb)	Allelic frequency		
		Siberia		Ural
		Krasnoyarsk Krai	Tomsk region	Bashkortostan
Ap049	118	0.005		
	121	0.005	0.003	
	123			0.917
	127	0.810	0.711	
	130	0.138	0.249	0.063
	138			0.021
	139	0.014	0.037	
	152	0.029		
Number of studied samples		**105**	**175**	**48**
Ap243	254			0.646
	255	0.280	0.524	
	257			0.354
	263	0.542	0.254	
	269	0.140	0.056	
	272	0.037	0.143	
	275		0.024	
Number of studied samples		**107**	**63**	**48**
H110	160			0.615
	162	0.624	0.837	
	163			0.302
	166	0.376	0.056	
	168			0.083
	170		0.107	
Number of studied samples		**117**	**135**	**48**

N, number of studied samples.

The populations in Siberia, specifically the Tomsk region and Krasnoyarsk Krai, have the most similar range and frequencies of alleles for the loci that have been analyzed (A008, Ap049, A113, Ap243, H110). The population in the Ural area, which is situated to the west of Siberia, exhibits variations from Siberia in some genetic loci. Specifically, differences were observed in the allele spectrum for locus

A008, in the allele frequency for locus A024, and in both the allele spectrum and frequency for loci Ap049 and Ap243. The Ural population exhibits a noteworthy resemblance to European populations in terms of the spectrum of alleles found at loci A024 and A008.

The study uncovered variations in the range of genetic variations and the prevalence of specific genetic variations at locus A008 among honeybees from Siberia, the Ural region, and European populations. The honeybees in the Ural and Europe had shorter alleles of locus A008 as the most common ones, with lengths of 154 bp and 148 bp respectively. On the other hand, bees from Siberia had allele "162" as the most distinct. This locus likely serves as a marker associated with geographic and environmental factors, indicating unique adaptation to local settings [1,3,48,49]. The study investigated various populations of dark-colored forest bees, including European, Ural, and Siberian populations. At certain genetic locations, such as A113, there is an overlap in the range of alleles present. However, the frequency of these alleles varies among various populations of honeybees. Various variables of population dynamics, such as the founder effect, genetic drift, and natural selection, can contribute to the occurrence of this phenomena.

Therefore, it has been demonstrated that certain loci exhibit distinct distributions of genotypes and alleles among bees, which vary based on their geographical location. Additional investigation is required, and the broadening of gene-geographic analyses of honeybees is pertinent. In order to evaluate the level of information provided by the examined genetic markers in distinguishing between distinct subspecies of honeybees, we compared the range of most common genetic variants in bees from different evolutionary branches (M and C) and from various geographical locations. The results of this comparison may be found in Table 7. An analysis of the variability of microsatellite loci in bees from various origins and geographical locations enables us to draw conclusions and make changes about the effectiveness of these loci as markers for distinguishing subspecies of honeybees. The spectrum of the most prevalent alleles at locus A008 varies among Apis m. mellifera populations in different geographical regions, as well as between the two southern races, Apis m. caucasica and Apis m. carpatica. No significant variations in the length of the most commonly observed allele were found in locus A113, neither among bees from the same lineage nor across bees of different races. This locus is likely not relevant for defining the subspecies. Loci A024 and Ap049 are highly intriguing for further investigation as potential markers to be included in the diagnostic panel, which may effectively distinguish between subspecies. In general, the locus A024 shows that the majority of bees and bee colonies Apis m. mellifera, regardless of their habitat, have shorter alleles. It is possible that there are variances in the allelic spectrum of locus Ap049 among bees of different races.

Table 7. Comparative analysis of the frequency of the most common alleles[*] of microsatellite loci in honeybees of different maternal origins and geographic localization.

	Geographical location	Sequence composition of the COI-COII mtDNA locus (breed)	Predominant allele	Allelic frequency
Locus A008	Tomsk region	PQQ/PQQQ	162	0.71–1.00
	Krasnoyarsky Krai	PQQ	162	1.00
	Ural (Bashkir population)[1]	PQQ	154	0.63–1.00
	Tomsk region[2]	Q	174	0.58–0.61
	Sochi area[3]	Q	158	0.88–1.00
	Europe[4]	*A.m.mellifera*	148	0.27–0.97
Locus A113	Tomsk region	PQQ/PQQQ	218 212 220	0.67–0.82 0.61 0.50
	Krasnoyarsky Krai	PQQ	218	0.85–0.95
	Ural (Bashkir population)[1]	PQQ	218 220	0.50–1.00 0.50
	Tomsk region[2]	Q	212	0.94–1.00
	Sochi area[3]	Q	222	0.50
	Europe[4]	*A.m.mellifera*	220	0.433–0.857
Locus A024	Tomsk region	PQQ/PQQQ	94 102	0.60–0.90 0.54
	Krasnoyarsky Krai	PQQ	98 96	0.50 0.50–0.71
	Ural (Bashkir population)[1]	PQQ	98 106	0.50–1.00 0.50
	Tomsk region[2]	Q	104	0.65
	Sochi area[3]	Q	106	0.88–1.00
	Europe[4]	*A.m.mellifera*	98	>0.80
Locus AP049	Tomsk region	PQQ/PQQQ	127 130	0.62–0.92 0.77
	Krasnoyarsky Krai	PQQ	127	0.50–0.96
	Ural (Bashkir population)[1]	PQQ	129 130	0.50–1.00 1.00
	Tomsk region[2]	Q	139	0.66–1.00
	Sochi area[3]	Q	139	1.00

[*]Data on allelic frequencies, the frequency of which = or > 0.5 are shown.

[1]Data on the Ural (Bashkir population) are taken from reference [47].

[2]Our own data for the Carpathian breed (*Apis m. carpatica*) imported into the territory of the Tomsk region from Carpathian breed nursery (d. Mukachevo, Ukraine).

[3]Data on the Caucasian honeybee (*Apis m. caucasica*) from the Sochi area are taken from reference [47].

[4]Data on the European population are taken from references [15,16].

To ascertain the subspecies classification of a honeybee individual, colony, or population, it is crucial to compare the numbers and genetic makeup of alleles across several research investigations. Unfortunately, there is currently no standardized reference material, such as a uniform allelic ladder, that is accessible for honeybees [3]. Furthermore, the spectrums of examined microsatellite markers frequently do not intersect, and key data regarding the allele spectrum and allele frequencies are not consistently provided in publications. Overall, the current phase of studying the variability of microsatellite loci in Apis mellifera can be seen as a period of gathering information. At this point in the honeybee study, it is advisable to exercise caution when using autosomal loci to identify the subspecies of honeybees.

3.5. Infestation of honeybees by *Nosema* in Siberia

The introduction of races of southern origin into the Siberian realm, where the Middle Russian breed had long been established, resulted in extensive crossbreeding of bees, a decline in purebred populations, reduced immunity, and an increased prevalence of diseases among bees. Conversely, the introduction of bee colonies from different regions (such as the European part of Russia and Uzbekistan) has had a negative impact on the epidemiological status, resulting in the development of hitherto unrecorded diseases in Siberia. An assessment was conducted to examine the prevalence of nosemosis: a comprehensive analysis was carried out to research the spread of Nosema infection across Siberia, identify the specific microsporidia species involved, and explore the source of bee colonies infected with Nosema. Nosemosis is a parasite disease that affects adult honeybees (Apis mellifera L.). It is caused by two known species of microsporidia, namely Nosema apis [50] and Nosema ceranae [51]. This illness is globally prevalent and has a substantial negative impact on honey production, leading to economic losses. The initial hypothesis posited that N. apis only targets the European honeybee species A. mellifera, resulting in nosemosis, while N. ceranae is a distinct disease that affects the Asian honeybee species A. cerana. It has recently been shown that N. ceranae is widely present throughout the global A. mellifera population, with confirmed cases in North and South America, Europe, and Asia [52–58]. The presence of the virus has been identified in Canada and the United States [59,60], and its existence has been verified in Central America [61], Australia [62], and North Africa [63]. The precise geographic distribution of Nosema in Russia remains poorly understood [64,65]. Furthermore, the available data on the occurrence of N. ceranae in Russia, particularly Siberia, is insufficient [66]. In the past, the occurrence of nosemosis in honeybees in Siberia was solely linked to N. apis. The issue of Nosema dispersion and its impact on honeybees remains unresolved. The impact of the Nosema infection on the longevity and efficiency of honeybees has not been well investigated. A screening research was conducted from 2012 to 2015 to examine the prevalence of Nosema spores in 132 bee colonies from 68 apiaries in Siberia [65]. In order to conduct an unbiased assessment, two distinct techniques were employed: microscopy and PCR. Our analysis revealed that 25.0% of the honeybees from 33 colonies out of the total 132 analyzed had spores that were identifiable using light microscopy. Additionally, 30.9% of the 21 apiaries out of the total 68 studied also had detectable spores. To overcome the challenge of morphologically distinguishing between N. ceranae and N. apis, a PCR technique utilizing 16S ribosomal RNA has been employed to discriminate the two species. In order to determine the exact species of Nosema present, we conducted PCR using primers that are unique to either N. apis or N. ceranae. Using optical microscopy, we examined spores from adult worker bees in 33 bee colonies (from 21 apiaries) to identify Nosema-positive samples. We next used PCR primers specific to the 16S rRNA gene to determine the species of Nosema, either N. ceranae or N. apis. Out of the 33 infected bee colonies from 19 apiaries, 28 colonies (84.8%) tested positive for N. apis using PCR with particular primers. Additionally, three colonies (9.1%) tested positive for N. ceranae, but this was only observed in two of the apiaries. Two bee colonies (2/33, 6.0%) from two apiaries were found to have samples that were co-infected with both N. ceranae and N. apis. The DNA fragments were sequenced to validate the PCR findings. The sequence analysis showed that N. apis (GenBank Accession No U97150) and N. ceranae (GenBank Accession No DQ486027) have identical sequences. Bees infected with Nosema were discovered in samples taken from five districts, primarily in the southern climatic regions of Siberia (specifically, the temperate continental sections) (see Figure 5). Bees infected with Nosema apis are imported from Uzbekistan in the northern area (Figure 5, C - Chainsky). The presence of Nosema ceranae in bees from the southern districts of the Tomsk region (Figure 5, Shegarsky and Tomsky districts) was determined to be the result of infected bees being brought in from southern parts of Russia.

The bees examined from apiaries in Krasnoyarsk Krai and Altai Krai showed no signs of Nosema infection. There is conflicting information in reports about how N. ceranae infections affect the health of honeybees and the survival of their colonies. Different signs of the disease have been documented in several studies [4,48,52,54–56,59,60,67–76]. Adult bees contract Nosema infections by consuming Nosema spores, which then develop in the midgut and invade the cells of the midgut epithelium. The predominant sign of Nosema infection produced by N. apis is diarrhea, but N. ceranae is known to cause mortality of people and colonies without any obvious symptoms preceding it [9,68]. The infection of Nosema apis is limited to the midgut epithelium [77], but N. ceranae has been identified using molecular techniques in other bee organs such as malpighian tubules and hypopharyngeal glands [78].

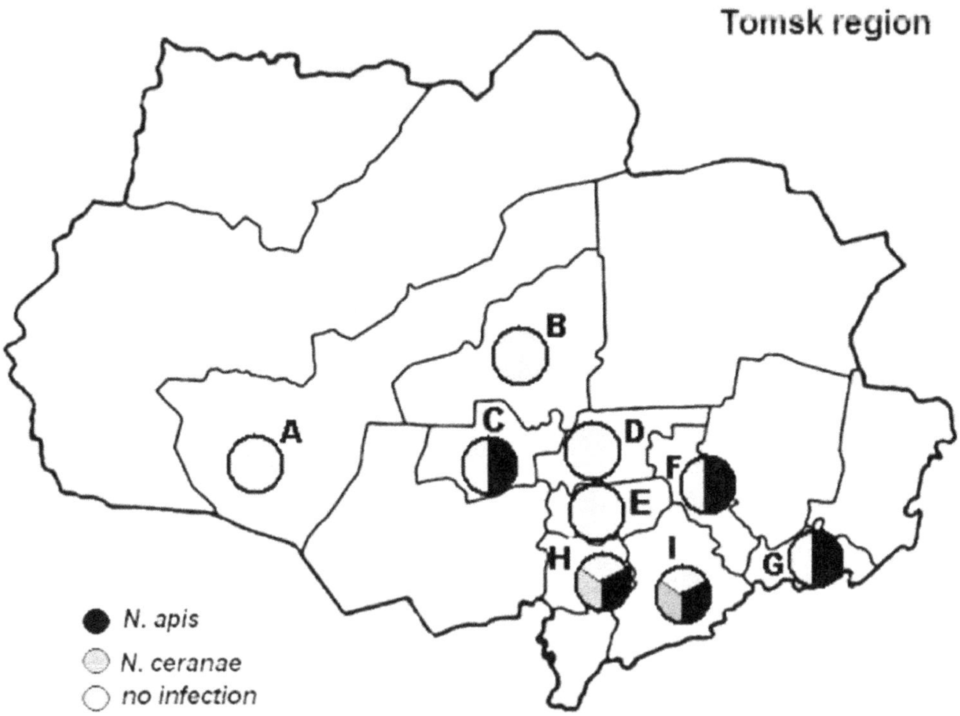

Figure 5. Distribution of Nosema in the honeybee colonies (Apis mellifera L.) throughout the Tomsk region (Western Siberia) (dots A–I). Bee colonies not infected by Nosema are indicated in yellow; bee colonies corresponding to infection by N. apis or N. ceranae are indicated in black and green, respectively. Sectors in circles indicate representation cases (existence/absence) of an infection without frequency. Literas (A–I) indicate the districts of the Tomsk region: A, Parabelsky; B, Kolpashevsky; C, Chainsky; D, Molchanovsky; E, Krivosheinsky; F, Asinovsky; G, Zyryansky; H, Shegarsky; I, Tomsky.

N. ceranae is likely the more dominant of the two Nosema species in terms of aggression towards the host, and it seems to be displacing N. apis in certain honeybee populations. The literature now discusses numerous factors contributing to the global prevalence of the parasite N. ceranae and its displacement of N. apis. Nosemosis caused by N. ceranae is a global issue due to the widespread presence of this parasite in many hosts [79]. Nosema ceranae exhibits greater parasitic aggression than N. apis, resulting in a wider distribution of N. ceranae compared to N. apis. However, the destruction of honeybee colonies by N. ceranae may be limited to specific regions rather than occurring worldwide [80]. The severity of N. ceranae's harmful effects may be affected by climate conditions [81-85] or may vary depending on the specific race and genetic diversity of honeybees [4,48,74,75,86-88].

The extent of infestation in honeybees is believed to be correlated with the race and origin (local or non-local) of the bees. Russian bee breeds have exhibited variations in their resistance to Nosema, as demonstrated in a study [86]. The infestation levels of N. ceranae varied significantly among lineages and colonies of both Russian and Italian workers [87]. Contrary to genetically similar Italian lines [87], Russian bee lineages exhibit significant genetic variation and are known for their strong resilience to diseases. There were considerable disparities in infection rates between bee colonies that were native to the area and those that were brought in from elsewhere [4,74]. Utilizing local honeybees increases the likelihood of colony survival due to their ability to adapt to specific regional environmental conditions, including climate and vegetation [48,75].

In order to ascertain the correlation between the occurrence of Nosema infection in bees and the different bee races, we conducted an analysis on bee breeds obtained from colonies that tested positive for Nosema. This analysis involved the use of morphometric measurements of wing venation and molecular-genetic analysis of mtDNA. The results of this analysis can be found in Table 8. The findings of the molecular genetic investigation, specifically targeting the COI-COII locus, of honeybees have been documented in reference [11]. The mtDNA study revealed the presence of PQQ and PQQQ variants of the COI-COII locus (A. m. mellifera, evolutionary branch M) in two colonies (families No. 2 and 4). Additionally, the Q variation (evolutionary branch C) was found in four colonies (families No. 1, 3, 5, and 7). Family No. 6 possessed bees with distinct variations of the COI-COII locus (PQQ and Q) and seemed to be the result of merging two colonies with diverse origins. Based on morphometric research, it has been determined that colonies No. 1, 3, 4, and 6 can be classified as subspecies of A. m. mellifera, while colonies No. 2, 5, and 7 are identified as hybrids. Based on the study of both morphometric and mtDNA data, it was found that only family No. 4 may be classified as A. m. mellifera. However, six bee colonies infected with Nosema did not match any of the established standards and were identified as honeybee hybrids (Table 8). In addition, several colonies that were observed had variations not only in morphometric parameters as compared to the established standards, but also in a discrepancy between the morphometric data and the results of the mtDNA study for the two honeybee colonies. Honeybees from colony No. 3 infected with N. apis and bees from colony No. 1 infected with N. ceranae belong to the A. m. mellifera race (branch M) based on the morphometric study. However, the mtDNA analysis indicated that these bees originated from branch C. Therefore, our findings suggest that the honeybees we studied, which were infected with Nosema, may represent a combination of two different races: Apis m. mellifera and Apis m. carpatica.

Table 8. Characterization honeybee colonies infested by *Nosema.**

№ colonies	*Nosema* species	Sequence composition of the COI-COII mtDNA locus	Morphometric parameters			
			Cubital index, standard units		Hantel index, standard units	
			Lim: min max	M ± m	Lim: min max	M ± m
1	*N. ceranae*	Q	1.30 2.29	1.66 ± 0.04	0.735 0.965	0.878 ± 0.011
2	*N. ceranae*	PQQQ	1.74 3.29	2.14 ± 0.07	0.857 1.053	0.937 ± 0.010
3	*N. apis*	Q	1.35 2.11	1.70 ± 0.03	0.667 0.917	0.804 ± 0.011
4	*N. apis*	PQQ	1.45 2.80	1.78 ± 0.06	0.754 1.0	0.846 ± 0.013
5	*N. apis*	Q	1.41 2.82	1.90 ± 0.06	0.656 1.176	0.880 ± 0.018
6	*N. apis*	PQQ/Q	1.28 2.80	1.73 ± 0.06	0.707 1.0	0.834 ± 0.015
7	*N. apis*	Q	1.43 2.35	1.86 ± 0.04	0.733 1.057	0.885 ± 0.011
Standard breeds (subspecies)**						
A. m. mellifera		PQQ, PQQQ and other	1.3 2.1	1.7	0.600 0.923	No data
A. m. carpatica		Q	2.3 3.0	2.65	≥0.925	No data

M ± m, average value of the sign ± the standard error of the mean.

*Thirty samples of bees were examined in each family.

**Definition of subspecies was carried out based on European standard honeybee [12].

To compare, we conducted a morphometric and mtDNA study to assess the origin of the bee colonies that were not infected with Nosema. This analysis was performed on 24 out of the 38 families that were investigated. Out of the 24 bee colonies that were not infected with Nosema, 18 colonies were identified as A. m. mellifera (75.0%), 3 colonies were identified as A. m. carpatica (12.5%), and 3 colonies were identified as hybrids (12.5%).

Currently, the frigid climate is seen as a significant constraint on the distribution of N. ceranae. The global spread of N. ceranae seems to be limited in colder areas [81,82], as N. ceranae spores can withstand high temperatures (60 °C) and drying, but they are not able to tolerate cold temperatures (4 °C) [81,82,89].

Even a brief exposure to low temperatures (4 °C) resulted in a significant decrease in N. ceranae spore germination, as observed in a study [82]. In regions with higher temperatures, N. ceranae outperforms N. apis in terms of competitiveness [48,82]. However, the spores of N. ceranae are significantly more susceptible to freezing compared to the spores of N. apis. The replacement of N. apis with N. ceranae is still a mystery [83].

The varying occurrence of N. ceranae can be attributed to its timing of introduction, either through natural dispersion or the introduction of infected honeybees, as well as the movement of bees within a country. The reduced or hindered germination of N. ceranae spores at low temperatures is expected to hinder the infectivity and dissemination of this pathogen in countries with cold winter seasons [82]. Our previous reports [65,66] have confirmed the extensive prevalence of N. ceranae in the Tomsk region of Western Siberia, Russia, indicating a global infection of N. ceranae in honeybee populations. Nevertheless, we discovered bee colonies infected with N. ceranae in regions characterized by cold climates, extended winters, and humid summers. It is worth noting that this parasite is not linked to colony depopulation or the collapse of honeybee populations. It was determined that these colonies, which had been infected before, were brought in from different regions of Russia. The presence of N. ceranae in Siberia, despite its harsh climate, contradicts the information suggesting that spores have low survival rates in cold temperatures. Simultaneously, the colonies affected by Nosema infection (either N. apis or N. ceranae) are primarily located in the southern regions of the Tomsk region. These areas are known for their advanced beekeeping practices and frequent introduction of southern bee breeds (A. m. caucasica and A. m. carpatica), resulting in extensive hybridization of honeybees. Introgressive hybridization alters the genetic composition of honeybee populations in a certain area, resulting in the erosion of their distinct genetic characteristics [4]. Hybridizing various subspecies of honeybees can result in the breakdown of gene complexes that have evolved over time. This can lead to a decline in the adaptive traits of animals and populations, as well as changes in the important biological and economic aspects of honeybees. The extensive hybridization of honeybees and subsequent generation of hybrid bees would undoubtedly facilitate the dissemination of diseases.

Our investigation found that most bee colonies infected with Nosema were of hybrid origin. This discovery aligns with the perspective that hybrid forms have limited ability to adjust to shifting environmental conditions and are more susceptible to disease. Hence, the outcomes of our investigation on the Nosema infestation in bee colonies are to be expected. Simultaneously, drawing a conclusion regarding the pathogenicity of a parasite from our data is unattainable. Hybrids may exhibit different developing circumstances of the parasite compared to pure breeds, which do not recognize the pathogenicity of N. ceranae in the host. Furthermore, there is an unresolved inquiry regarding the spread of a Nosema infection in the northern area of the Tomsk region. This inquiry investigates the impact of a cold climate and a limited number of hybrid bees on the distribution of Nosema. It is worth noting that no colonies infected with Nosema have been found in this region, except in the Chainsky district where bees infected with N. apis were imported from Uzbekistan. Siberia offers an optimal setting for examining the relationship between the spread of this disease and climatic factors, as well as understanding its transmission to extremely isolated regions. This idea is particularly fascinating because the impact of climate change on illness prevalence and the strength of disease-causing agents is a topic of extensive discussion [80]. Clearly, further investigation is required to clarify the complete impact of N. ceranae infection on A. mellifera colonies in various geographic regions, and to determine whether there are variations in the virulence levels of individual parasites and colony virulence levels between the two.

4. CONCISE SUMMARY

This study conducted in Siberia highlights the necessity of adopting a comprehensive approach to investigate many aspects of honeybees. These elements include distinguishing between subspecies, understanding the influence of environmental (geographical) factors on the development of genetic diversity in bees, and examining the prevalence of bee diseases. The main objective of studying the genetic diversity of honeybees is to ascertain their subspecies composition. When doing gene-geographical research, it is crucial to evaluate the adaptive and selective importance of genetic markers. This is also crucial for the strategic organization and execution of projects with practical applications. In addition to traditional external traits that have long been used to distinguish honeybee breeds, molecular genetic approaches are also widely utilized. Nevertheless, due to the significant hybridization of bees, where approximately one-third of bee colonies exhibit a disparity between genetic and morphometric characteristics, and occasionally experience a complete mismatch, a thorough examination of the bees is required. The occurrence of hybrid forms in a region under study for genetic diversity has two negative consequences. Firstly, it hinders the preservation of gene pools of distinct subspecies, such as the dark-colored forest bee. Secondly, it complicates the search for traits that are both adaptive and economically valuable, potentially leading to distorted results and their interpretation. Hence, it is imperative to consider this factor when carrying out such research. The data above suggest that relying solely on external or genetic traits may not be enough to determine the origin of bees. To accurately assess the breed and cases of hybridization, it is necessary to simultaneously analyze morphometric parameters and data on the variability of the COI-COII locus of mtDNA. Given the prevalence of bee crossbreeding, it is imperative to enhance genetic techniques for maintaining the integrity of bee colonies. International and Russian scholars [43,47,90] conduct research in this field. Based on thorough research conducted in Eastern Europe, which involved searching for informative markers among over 1,000 SNP using five different analytical methods, five panels have been created. These panels consist of 48, 96, 144, 192, and 284 markers that are informative for determining the ancestral origin of species. The authors suggest utilizing the findings of this investigation to detect and assess the contamination of C-lines (namely, Apis m. ligustica and Apis m. carnica) with the M-line (Apis m. mellifera) [43]. Initial studies conducted by Russian researchers indicate that honeybee populations in Russia exhibit significant genetic diversity. However, it is unlikely that a standardized set of genetic markers can be developed to differentiate between different bee breeds across the entire country. Integrating the scientific achievements and results of many laboratories and scientific groups worldwide is essential to identify universal patterns of genetic diversity in bees and evaluate the adaptive and selective capabilities of honeybees.

REFERENCES

1. De la Rúa P, Jaffé R, Dall'Olio R, Muñoz I, Serrano J. Biodiversity, conservation and current threats to European honey bees. Apidologie. 2009;40(3):263-284. DOI: 10.1051/apido/2009027

2. Meixner MD, Costa C, Kryger P, Hatjina F, Bouga M, Ivanova E, Büchler R. Conserving diversity and vitality for honey bee breeding. Journal of Apicultural Research. 2010;49(1):85-92. DOI: 10. 3896/IBRA.1.49.1.12

3. Meixner MD, Pinto MA, Bouga M, Kryger P, Ivanova E, Fuchs S. Standard methods for characterising subspecies and ecotypes of Apis mellifera. In: Dietemann V, Ellis JD, Neumann P, Editors. The COLOSS BEEBOOK, Volume I: Standard Methods for Apis mellifera Research. Journal of Apicultural Research. 2013;52(4):1-28. DOI 10.3896/IBRA.1.52.4.05

4. Büchler R, Costa C, Hatjina F, Andonov S, Meixner MD, Le Conte Y, Uzunov A, Berg S, Bienkowska M, Bouga M, Drazic M, Dyrba W, Kryger P, Panasiuk B, Pechhacker H, Petrov P, Kezić N, Korpela S, Wilde J. The influence of genetic origin and its interaction with environmental effects on the survival of Apis mellifera L. colonies in Europe. Journal of Apicultural Research. 2014;53(2):205-214. DOI: 10.3896/IBRA.1.53.2.03

5. Nikonorov YM, Ben'kovskaya GV, Poskryakov AV, Nikolenko AG, Vakhitov VA. The use of the PCR technique for control of the pure-breeding of honeybee

(Apis mellifera mellifera L.) colonies from the Southern Urals. Russian Journal of Genetics. 1998;34(11):1344-1347.

6. Ostroverkhova NV, Konusova OL, Kucher AN, Pogorelov YL, Belykh EA, Vorotov AA. Population genetic structure of honey bee (Apis mellifera L.) in the village of Leboter in Chainsky district of the Tomsk region. Tomsk State University Journal of Biology. 2013;1(21):161-172.

7. Solignac M, Vautrin D, Loiseau A, Mougel F, Baudry E. Five hundred and fifty microsatellite markers for the study of the honey bee (Apis mellifera L.) genome. Molecular Ecology Notes. 2003;3:307-311. DOI: 10.1046/j.1471-8286.2003.00436.x

8. Fries I, Chauzat MP, Chen YP, Doublet V, Genersch E, Gisder S, Higes M, McMahon DP, Martín-Hernández R, Natsopoulou M, Paxton RJ, Tanner G, Webster TC, Williams GR. Standard methods for Nosema research. In: Dietemann V, Ellis JD, Neumann P, Editors. The COLOSS BEEBOOK, Volume II: Standard Methods for Apis mellifera Pest and Pathogen Research. Journal of Apicultural Research. 2013;52(1):1-28. DOI: 10.3896/IBRA.1.52.1.14

9. Martín-Hernández R, Meana A, Prieto L, Salvador AM, Garrido-Bailon E, Higes M. Outcome of colonization of Apis mellifera by Nosema ceranae. Applied and Environmental Microbiology. 2007;73(20):6331-6338. DOI: 10.1128/aem.00270-07

10. Hamiduzzaman MM, Guzman-Novoa E, Goodwin PH. A multiplex PCR assay to diagnose and quantify Nosema infections in honey bees (Apis mellifera). Journal of Invertebrate Pathology. 2010;105(2):151-155. DOI: 10.1016/j.jip.2010.06.001

11. Ostroverkhova NV, Konusova OL, Kucher AN, Kireeva TN, Vorotov AA, Belikh EA. Genetic diversity of the locus COI-COII of mitochondrial DNA in honeybee populations (Apis mellifera L.) from the Tomsk region. Russian Journal of Genetics. 2015;51(1):80-90. DOI: 10.1134/S102279541501010X

12. Cauia E, Usurelu D, Magdalena LM, Cimponeriu D, Apostol P, Siceanu A, Holban A, Gavrila L. Preliminary researches regarding the genetic and morphometric characterization of honeybee (A. mellifera L.) from Romania. Scientific Papers Animal Science and Biotechnologies. 2008;41(2):278-286.

13. Estoup A, Solignac M, Harry M, Cornuet JM. Characterization of (GT), and (CT) microsatellites in two insect species: Apis mellifera and Bombus terrestris. Nucleic Acids Research. 1993;21:1427-1431.

14. Estoup A, Garnery L, Solignac M, Cornuet JM. Microsatellite variation in honey bee (Apis mellifera L.) populations: Hierarchical genetic structure and test of the infinite allele and stepwise mutation models. Genetics. 1995;140:679-695.

15. Franck P, Garnery L, Solignac M, Cornuet JM. The origin of West European subspecies of honeybees (Apis mellifera): New insights from microsatellite and mitochondrial data. Evolution. 1998;52(4):1119-1134. DOI: 10.2307/2411242

16. Garnery L, Franck P, Baudry E, Vautrin D, Cornuet JM, Solignac M. Genetic diversity of the West European honey bee (Apis mellifera mellifera and A. m. iberica). II. Microsatellite loci. Genetics Selection and Evolution. 1998;30(1):S49-S74.

17. Franck P, Garnery L, Solignac M, Cornuet JM. Molecular confirmation of a fourth lineage in honeybees from the Near East. Apidologie. 2000;31(2):167-180. DOI: 10.1051/apido:2000114

18. Franck P, Garnery L, Loiseau A, Oldroyd BP, Hepbum HR, Solignac M, Cornuet JM. Genetic diversity of the honeybee in Africa: microsatellite and mitochondrial data. Heredity. 2001;86(4):420-430.

19. De la Rúa P, Galián J, Serrano J, Moritz RFA. Microsatellite analysis of non-migratory colonies of Apis mellifera iberica from South-eastern Spain. Journal of Zoological Systematics and Evolutionary Research. 2002;40(3):164-168. DOI: 10.1046/j.1439-0469.2002.00187.x

20. De la Rúa P, Galián J, Serrano J, Moritz RFA. Genetic structure of Balearic honeybee populations based on microsatellite polymorphism. Genetics Selection Evolution. 2003;35:339-350. DOI: 10.1051/gse:2003012

21. De la Rúa P, Jimenez Y, Galián J, Serrano J. Evaluation of the biodiversity of honey bee (Apis mellifera) populations from Eastern Spain. Journal of Apicultural Research. 2004;43(4):162-166. DOI: 10.1080/00218839.2004.11101130

22. Sušnik S, Kozmus P, Poklukar J, Meglic V. Molecular characterisation of indigenous Apis mellifera carnica in Slovenia. Apidologie. 2004;35(6):623-636. DOI: 10.1051/apido:2004061

23. Jensen AB, Palmer KA, Boomsma JJ, Pedersen BoV. Varying degrees of Apis mellifera ligustica introgression in protected populations of the black honeybee, Apis mellifera mellifera, in Northwest Europe. Molecular Ecology. 2005;14(1):93-106. DOI: 10.1111/j.1365-294X.2004.02399.x

24. De la Rúa P, Galián J, Pedersen BoV, Serrano J. Molecular characterization and population structure of Apis mellifera from Madeira and the Azores. Apidologie. 2006;37(6):699-708. DOI: 10.1051/apido:2006044

25. Kandemir I, Meixner MD, Ozkan A, Sheppard WS. Genetic characterization of honey bee (Apis mellifera cypria) populations in northern Cyprus. Apidologie. 2006;37(5):547-555. DOI: 10.1051/apido:2006029

26. Bodur C, Kence M, Kence A. Genetic structure of honeybee, Apis mellifera L. (Hymenoptera: Apidae) populations of Turkey inferred from microsatellite

analysis. Journal of Apicultural Research. 2007;46(1):50-56. DOI: 10.3896/IBRA.1.46.1.09

27. Dall'Olio R, Marino A, Lodesani M, Moritz RFA. Genetic characterization of Italian honeybees, Apis mellifera ligustica, based on microsatellite DNA polymorphisms. Apidologie. 2007;38(2):207-217. DOI: 10.1051/apido:2006073
28. Lattorff HM, Moritz RFA, Crewe RM, Solignac M. Control of reproductive dominance by the thelytoky gene in honeybees. Biology letters. 2007;3(3):292-295. DOI: 10.1098/rsbl.2007.0083
29. Miguel I, Iriondo M, Garnery L, Sheppard WS, Estonba A. Gene flow within the M evolutionary lineage of Apis mellifera: role of the Pyrenees, isolation by distance and post-glacial re-colonization routes in the Western Europe. Apidologie. 2007;38(2):141-155. DOI: 10.1051/apido:2007007
30. Solignac M, Mougel F, Vautrin D, Monnerot M, Cornuet JM. A third-generation microsatellite-based linkage map of the honey bee, Apis mellifera, and its comparison with the sequence-based physical map. Genome Biology. 2007;8:R66. DOI: 10.1186/gb-2007-8-4-r66
31. Bourgeois L, Sylvester A, Danka R, Rinderer T. Comparison of microsatellite DNA diversity among commercial queen breeder stocks of Italian honey bees in the United States and Italy. Journal of Apicultural Research and Bee World. 2008;47(2):93-98. DOI: 10.3896/IBRA.1.47.2.01
32. Moritz RFA, Dietemann V, Crewe R. Determining colony densities in wild honeybee populations (Apis mellifera) with linked microsatellite DNA markers. Journal of Insect Conservation. 2008;12(5):455-459. DOI: 10.1007/s10841-007-9078-5
33. Hernández-García R, De la Rúa P, Serrano J. Mating frequency in Apis mellifera iberiensis queens. Journal of Apicultural Research. 2009;48(2):121-125. DOI: 10.3896/IBRA.1.48.2.06
34. Muñoz I, Dall'Olio R, Lodesani M, De la Rúa P. Population genetic structure of coastal Croatian honey bees (Apis mellifera carnica). Apidologie. 2009;40(6):617-626. DOI: 10.1051/apido/2009041
35. Soland-Reckeweg G, Heckel G, Neumann P, Excoffier L. Gene flow in admixed populations and implications for the conservation of the Western honeybee, Apis mellifera. Journal of Insect Conservation. 2009;13(3):317-328. DOI: 10.1007/s10841-008-9175-0
36. Miguel I, Baylac M, Iriondo M, Manzano C, Garnery L, Estonba A. Both geometric morphometric and microsatellite data consistently support the differentiation of the Apis mellifera M evolutionary branch. Apidologie. 2010; 42:150-161. DOI: 10.1051/apido/2010048

37. Canovas F, De la Rúa P, Serrano J, Galián J. Microsatellite variability reveals beekeeping influences on Iberian honeybee populations. Apidologie. 2011;42(3):235-251. DOI: 10.1007/s13592-011-0020-1

38. Nikolova SR. Genetic variability of local Bulgarian honey bees Apis mellifera macedonica (rodopica) based on microsatellite DNA analysis. Journal of Apicultural Science. 2011;55(2):117-129.

39. Oleksa A, Chybicki I, Tofilski A, Burczyk J. Nuclear and mitochondrial patterns of introgression into native dark bees (Apis mellifera mellifera) in Poland. Journal of Apicultural Research. 2011;50(2):116-129. DOI: 10.3896/IBRA.1.50.2.03

40. Muñoz I, De la Rúa P. Temporal analysis of the genetic diversity in a honey bee mating area of an Island population (La Palma, Canary Islands, Spain). Journal of Apicultural Science. 2012;56(1):41-49. DOI: 10.2478/v10289-012-0005-y

41. Nedić N, Francis RM, Stanisavljević L, Pihler I, Kezić N, Bendixen C, Kryger P. Detecting population admixture in the honey bees of Serbia. Journal of Apicultural Research. 2014;53:303-313. DOI: 10.3896/ibra.1.53.2.12

42. Uzunov A, Costa C, Panasiuk B, Meixner MD, Kryger P, Hatjina F, Bouga M, Andonov S, Bienkowska M, Le Conte Y, Wilde J, Gerula D, Kiprijanovska H, Filipi J, Petrov P, Ruottinen L, Pechhacker H, Berg S, Dyrba W, Ivanova E, Buchler R. Swarming, defensive and hygienic behaviour in honey bee colonies of different genetic origin in a pan-European experiment. Journal of Apicultural Research. 2014;53:248-260. DOI: 10.3896/IBRA.1.53.2.06

43. Muñoz I, Henriques D, Johnston JS, Chávez-Galarza J, Kryger P, Pinto MA. Reduced SNP panels for genetic identification and introgression analysis in the dark honey bee (Apis mellifera mellifera). PLoS ONE. 2015;10(4):e0124365. DOI: 10.1371/journal.pone.0124365

44. Nikolova SR, Bienkowska M, Gerula D, Ivanova EN. Microsatellite DNA polymorphism in selectively controlled Apis mellifera carnica and Apis mellifera caucasica populations from Poland. Archives of Biological Sciences. 2015;67(3):889-894. DOI:10.2298/ABS141102048N

45. Techer MA, Clémencet J, Simiand C, Portlouis G, Reynaud B, Delatte H. Genetic diversity of the honeybee (Apis mellifera L.) populations in the Seychelles archipelago. Insect Conservation and Diversity. 2016;9(1):13-26. DOI: 10.1111/icad.12138

46. Ostroverkhova NV, Konusova OL, Kucher AN, Kireeva TN. Investigation of polyandry in honey bees (Apis mellifera) using microsatellites. Russian Journal of Zoology. 2016;95(3):307-313. DOI: 10.7868/S0044513416030119

47. Ilyasov RA, Kosarev MN, Poskryakov AV, Sharipov AY, Nikolenko AG. New approach to the assessment of genetic potential of colonies of dark European bee Apis mellifera mellifera based on polymorphism of microsatellite loci. Biomiks. 2015;7(2):138-152.

48. Meixner MD, Büchler R, Costa C, Francis RM, Hatjina F, Kryger P, Uzunov A, Carreck NL. Honey bee genotypes and the environment. Journal of Apicultural Research. 2014;53(2):183-187. DOI: 10.3896/IBRA.1.53.2.01

49. Hatjina F, Costa C, Büchler R, Uzunov A, Drazic M, Filipi J, Charistos L, Ruottinen L, Andonov S, Meixner MD, Bienkowska M, Dariusz G, Panasiuk B, Le Conte Y, Wilde J, Berg S, Bouga M, Dyrba W, Kiprijanovska H, Korpela S, Kryger P, Lodesani M, Pechhacker H, Petrov P, Kezic N. Population dynamics of European honey bee genotypes under different environmental conditions. Journal of Apicultural Research. 2014;53(2):233-247. DOI: 10.3896/IBRA.1.53.2.05

50. Zander E. Tierische Parasiten als Krankheitserreger bei der Biene. Münchener Bienenzeitung. 1909;31:196-204.

51. Fries I, Feng F, daSilva A, Slemenda SB, Pieniazek NJ. Nosema ceranae n. sp. (Microspora, Nosematidae), morphological and molecular characterization of a microsporidian parasite of the Asian honey bee Apis cerana (Hymenoptera, Apidae). European Journal of Protistology. 1996;32(3):356-365. DOI:10.1016/s0932-4739(96)80059-9

52. Higes M, Martin R, Meana A. Nosema ceranae, a new microsporidian parasite in honeybees in Europe. Journal of Invertebrate Pathology. 2006;92(2):93-95. DOI: 10.1016/j.jip.2006.02.005

53. Huang WF, Jiang JH, Chen YW, Wang CH. A Nosema ceranae isolate from the honeybee Apis mellifera. Apidologie. 2007;38:30-37. DOI: 10.1051/apido:2006054

54. Klee J, Besana AM, Genersch E, Gisder S, Nanetti A, Tam DQ, Chinh TX, Puerta F, Ruz JM, Kryger P, Message D, Hatjina F, Korpela S, Fries I, Paxton RJ. Widespread dispersal of the microsporidian Nosema ceranae, an emergent pathogen of the western honey bee, Apis mellifera. Journal of Invertebrate Pathology. 2007;96:1-10. DOI: 10.1016/j.jip.2007.02.014

55. Paxton RJ, Klee J, Korpela S, Fries I. Nosema ceranae has infected Apis mellifera in Europe since at least 1998 and may be more virulent than Nosema apis. Apidologie. 2007;38:558-565. DOI: 10.1051/apido:2007037

56. Invernizzi C, Abud C, Tomasco IH, Harriet J, Ramallo G, Campá J, Katz H, Gardiol G, Mendoza Y. Presence of Nosema ceranae in honeybees (Apis mellifera) in Uruguay. Journal of Invertebrate Pathology. 2009;101(2):150-153. DOI: 10.1016/j.jip.2009.03.006

57. Paxton RJ. Does infection by Nosema ceranae cause "Colony Collapse Disorder" in honey bees (Apis mellifera)? Journal of Apicultural Research. 2010;49(1):80-84. DOI: 10.3896/IBRA.1.49.1.11

58. Nabian C, Ahmadi K, Nazem Shirazi MH, Gerami Sadeghian A. First detection of Nosema ceranae, a microsporidian protozoa of European honeybees (Apis mellifera) in Iran. Iranian Journal of Parasitology. 2011;6(3):89-95.

59. Chen Y, Evans JD, Smith IB, Pettis JS. Nosema ceranae is a long-present and wide-spread microsporidian infection of the European honey bee (Apis mellifera) in the United States. Journal of Invertebrate Pathology. 2008;97:186-188. DOI: 10.1016/j.jip.2007.07.010

60. Williams GR, Shafer ABA, Rogers REL, Shutler D, Stewart DT. First detection of Nosema ceranae, a microsporidian parasite of European honey bees (Apis mellifera), in Canada and central USA. Journal of Invertebrate Pathology. 2008;97(2):189-192. DOI: 10.1016/j.jip.2007.08.005

61. Calderón RA, Sanchez LA, Yañez O, Fallas N. Presence of Nosema ceranae in Africanized honey bee colonies in Costa Rica. Journal of Apicultural Research. 2008;47:328-329. DOI: 10.3896/IBRA.1.47.4.18

62. Giersch T, Berg T, Galea F, Hornitzky M. Nosema ceranae infects honey bees (Apis mellifera) and contaminates honey in Australia. Apidologie. 2009;40:117-123. DOI: 10.1051/apido/2008065

63. Higes M, Martín-Hernández R, Garrido-Bailón E, Botias C, Meana A. The presence of Nosema ceranae (Microsporidia) in North African honey bees (Apis mellifera intermissa). Journal of Apicultural Research. 2009;48:217-219. DOI: 10.3896/ibra.1.48.3.12

64. Zinatullina ZJ, Zhigileva ON, Ignatjeva AN, Tokarev YS. Nosemosis of honeybees on the apiaries in Russian [Internet]. 2012. Available from: http://www.apiworld.ru/khochu-vsye-znat/iii-mezhdunarodnyy-forum-pchelovodov-medovyy-mir/nozematoz-na-pasekakh-rossii/ (Accessed 2015-12-12).

65. Ostroverkhova NV, Konusova OL, Kucher AN, Simakova AV, Golubeva EP, Kireeva TN, Sharakhov IV. Infestation of honeybee (Apis mellifera) by Nosema (Microsporidia) in the Tomsk region. Parasitology. 2016;50(3).

66. Ostroverkhova NV, Konusova OL, Pogorelov YL, Kireeva TN, Salik MY, Golubeva EP. The first case of detection of Nosema ceranae on the apiary of the Tomsk region. Pchelovodstvo. 2014;9:28-30.

67. Fries I, Martin R, Meana A, García-Palencia P, Higes M. Natural infections of Nosema ceranae in European honey bees. Journal of Apicultural Research. 2006;45:230-233. DOI: 10.3896/ibra.1.45.4.13

68. Higes M, Martín-Hernández R, Botías C, Garrido-Bailón E, González-Porto AV, Barrios L, del Nozal MJ, Bernal JL, Jiménez JJ, García-Palencia P, Meana A. How natural infection by Nosema ceranae causes honeybee colony collapse. Environmental Microbiology. 2008;10:2659-2669. DOI: 10.1111/j.1462-2920.2008.01687.x

69. Chauzat MP, Higes M, Martín-Hernández R, Meana A, Cougoule N, Faucon JP. Presence of Nosema ceranae in French honey bee colonies. Journal of Apicultural Research. 2007;46(2):127-128. DOI:10.1080/00218839.2007.11101380

70. Cox-Foster DL, Conlan S, Holmes EC, Palacios G, Evans JD, Moran NA, Quan P-L, Briese S, Hornig M, Geiser DM, Martinson V, VanEngelsdorp D, Kalkseitn AL, Drysdale L, Hui J, Zhai J, Cui L, Hutchison S, Simons JF, Egholm M, Pettis JS, Lipkin WI. A metagenomic survey of microbes in honey bee colony collapse disorder. Science. 2007;318:283-287. DOI: 10.1126/science.1146498

71. Mulholland GE, Traver BE, Johnson NG, Fell RD. Individual variability of Nosema ceranae infections in Apis mellifera colonies. Insects. 2012;3(4):1143-1155. DOI: 10.3390/insects3041143

72. Botias C, Martín-Hernández R, Barrios L, Meana A, Higes M. Nosema spp. infection and its negative effects on honey bees (Apis mellifera iberiensis) at the colony level. Veterinary Research. 2013;44(1):25. DOI: 10.1186/1297-9716-44-25

73. Goblirsch M, Huang ZY, Spivak M. Physiological and behavioral changes in honey bees (Apis mellifera) induced by Nosema ceranae infection. PLoS ONE. 2013;8(3):e58165. DOI: 10.1371/journal.pone.0058165

74. Francis RM, Amiri E, Meixner MD, Kryger P, Gajda A, Andonov S, Uzunov A, Topolska G, Charistos L, Costa C, Berg S, Bienkowska M, Bouga M, Büchler R, Dyrba W, Hatjina F, Ivanova E, Kezic N, Korpela S, Le Conte Y, Panasiuk B, Pechhacker H, Tsoktouridis G, Wilde J. Effect of genotype and environment on parasite and pathogen levels in one apiary – a case study. Journal of Apicultural Research. 2014;53(2):230-232. DOI: 10.3896/IBRA.1.53.2.14

75. Meixner MD, Francis RM, Gajda A, Kryger P, Andonov S, Uzunov A, Topolska G, Costa C, Amiri E, Berg S, Bienkowska M, Bouga M, Büchler R, Dyrba W, Gurgulova K, Hatjina F, Ivanova E, Janes M, Kezic N, Korpela S, Le Conte Y, Panasiuk B, Pechhacker H, Tsoktouridis G, Vaccari G, Wilde J. Occurrence of parasites and pathogens in honey bee colonies used in a European genotype-environment interactions experiment. Journal of Apicultural Research. 2014;53(2):215-229. DOI: 10.3896/IBRA.1.53.2.04

76. Milbrath MO, van Tran T, Huang WF, Solter LF, Tarpy DR, Lawrence F, Huang ZY. Comparative virulence and competition between Nosema apis and Nosema

ceranae in honey bees (Apis mellifera). Journal of Invertebrate Pathology. 2015;125:9-15. DOI: 10.1016/j.jip.2014.12.006

77. Fries I. Infectivity and multiplication of Nosema apis Z. in the ventriculus of the honey bee. Apidologie. 1988;19:319-328. DOI: 10.1051/apido:19880310

78. Chen YP, Evans JD, Murphy C, Gutell R, Zuker M, Gundensen-Rindal D, Pettis JS. Morphological, molecular, and phylogenetic characterization of Nosema ceranae, a microsporidian parasite isolated from the European honey bee, Apis mellifera. Journal of Eukaryotic Microbiology. 2009;56:142-147. DOI: 10.1111/j.1550-7408.2008.00374.x

79. Higes M, Meana A, Bartolomé C, Botías C, Martín-Hernández R. Nosema ceranae (Microsporida), a controversial 21st century honey bee pathogen. Environmental Microbiology Reports. 2013;5(1):17-29. DOI:10.1111/1758-2229.12024

80. Genersch E. Honey bee pathology: current threats to honey bees and beekeeping. Applied Microbiology and Biotechnology. 2010;87:87-97. DOI: 10.1007/s00253-010-2573-8

81. Fries I. Nosema ceranae in European honey bees (Apis mellifera). Journal of Invertebrate Pathology. 2010;103:S73-S79. DOI: 10.1016/j.jip.2009.06.017

82. Gisder S, Hedtke K, Möckel N, Frielitz MC, Linde A, Genersch E. Five-year cohort study of Nosema spp. in Germany: does climate shape virulence and assertiveness of Nosema ceranae? Applied and Environmental Microbiology. 2010;76(9):3032-3038. DOI: 10.1128/AEM.03097-09

83. Higes M, García-Palencia P, Botias C, Meana A, Martín-Hernández R. The differential development of microsporidia infecting worker honey bee (Apis mellifera) at increasing incubation temperature. Environmental Microbiology Reports. 2010;2(6):745-748. DOI: 10.1111/j.1758-2229.2010.00170.x.

84. Sánchez Collado JG, Higes M, Barrio L, Martín-Hernández R. Flow cytometry analysis of Nosema species to assess spore viability and longevity. Parasitology Research. 2014;113(5):1695-1701. DOI:10.1007/s00436-014-3814-z

85. Van der Zee R, Gómez-Moracho T, Pisa L, Sagastume S, García-Palencia P, Maside X, Bartolomé C, Martín-Hernández R, Higes M. Virulence and polar tube protein genetic diversity of Nosema ceranae (Microsporidia) field isolates from Northern and Southern Europe in honeybees (Apis mellifera iberiensis). Environmental Microbiology Reports. 2014;6(4):401-413. DOI: 10.1111/1758-2229.12133

86. Kharitonov NN. Breeding of honeybees that are resistant to disease. Pchelovodstvo. 2006;7:14-16.

87. Bourgeois AL, Rinderer TE, Sylvester HA, Holloway B, Oldroyd BP. Patterns of Apis mellifera infestation by Nosema ceranae support the parasite hypothesis for the evolution of extreme polyandry in eusocial insects. Apidologie. 2012;43(5):539-548. DOI: 10.1007/s13592-012-0121-5

88. Chauzat MP, Laddomada A. Foreword. In: Dietemann V, Ellis JD, Neumann P, Editors. The COLOSS BEEBOOK, Volume II: Standard Methods for Apis mellifera Pest and Pathogen Research. Journal of Apicultural Research. 2013;52(4):1-2. DOI: 10.3896/IBRA.1.52.4.17

89. Fenoy S, Rueda C, Higes M, Martín-Hernández R, del Aguila C. High-level resistance of Nosema ceranae, a parasite of the honeybee, to temperature and desiccation. Applied and Environmental Microbiology. 2009;75(21):6886-6889. DOI: 10.1128/AEM.01025-09

90. Ostroverkhova NV, Konusova OL, Kucher AN, Kireeva TN, Baghirov RT-o. Characterization of the genetic diversity of honey bees (Apis mellifera L.) in Tomsk population using mtDNA and microsatellite markers. A.I. Kurentsov's Annual Memorial Meetings. 2015;XXVI:227-240.

02 PRINCIPLES FOR DESIGNING A BREEDING PROGRAM FOR ROYAL JELLY

1. INTRODUCTION: AN OVERVIEW

In spite of the increasing honey export numbers in recent years, beekeeping in developing countries is experiencing a state of technological stagnation. This means that there has been no significant improvement in the genetic quality of honeybees, and no new efficient management techniques have been developed to enhance productivity. The strong demand for honeybee colonies with superior performance and acceptable behavioral traits leads to the alteration of natural biodiversity through the widespread importation of queens [1]. According to Almeida and Carvalho [2], in order to succeed in the highly competitive bee product market, beekeepers must focus on innovation in management and technology. They should approach beekeeping from a business perspective and also play a role in preserving the genetic diversity of honeybees. In Brazil, the federal government and service firms have not provided significant financial support or technical training to beekeepers, hindering the dissemination of new or old technology. Nevertheless, due to the depreciation of the national currency, the real, against the dollar in recent years, beekeeping has become increasingly lucrative and competitive compared to other agricultural pursuits. However, it is worth noting that while there has been a rise in the number of managed hives, there has not been a significant increase in the number of individuals engaged in beekeeping.

Genetic breeding is a method that can be employed to enhance production. Animal breeding programs have progressed over the past decades by utilizing advancements in science and technology to better identify and choose the most suitable individuals for breeding in order to improve the genetic information available for the following generation. Breeding programs utilize calculations, scientific principles, biotechnology, and advancements in computing and information technology to achieve a high level of process efficiency. Enhancing honeybee traits is crucial for beekeepers, and doing this requires integrating honeybee adaptation to the environment, productivity, and economic sustainability for beekeepers [4]. Kinghorn et al. [5] found that the use of breeding techniques is influenced by the interplay between technological feasibility and socio-economic acceptability within the production system.

The process of Africanization resulted in substantial modifications in the rearing and production practices, which, in conjunction with research efforts, facilitated advancements in artificial insemination, queen production, genetic breeding with the identification of strains for honey production, royal jelly, and hygienic behavior. Consequently, the desire for increased professionalism among beekeepers prompted the quest for knowledge and the implementation of genetic breeding programs.

One of the key aspects contributing to improvements in productivity is animal breeding. There are numerous programs in our country that are focused on tactics aimed at generating demonstrated genetic advancement in beekeeping. For example, by continuously producing chosen queens, beekeepers can promptly replace deceased or aging queens with young queens that possess desirable genetic and phenotypic traits. Furthermore, the optimal growth and efficiency of the colony mostly rely on the age and caliber of its queen [6], as the offspring inherit fifty percent of their traits from the maternal queen. Various factors, including genotype, diet, production methods, production timing, and larval age, influence the quality of the queens [7].

The organization of a breeding program entails the strategic considerations that are crucial for its achievement, particularly in the case of honeybees. The definition of the strategies is illustrated in Figure 1, and the training of personnel for their development is a crucial aspect in the program's success [8].

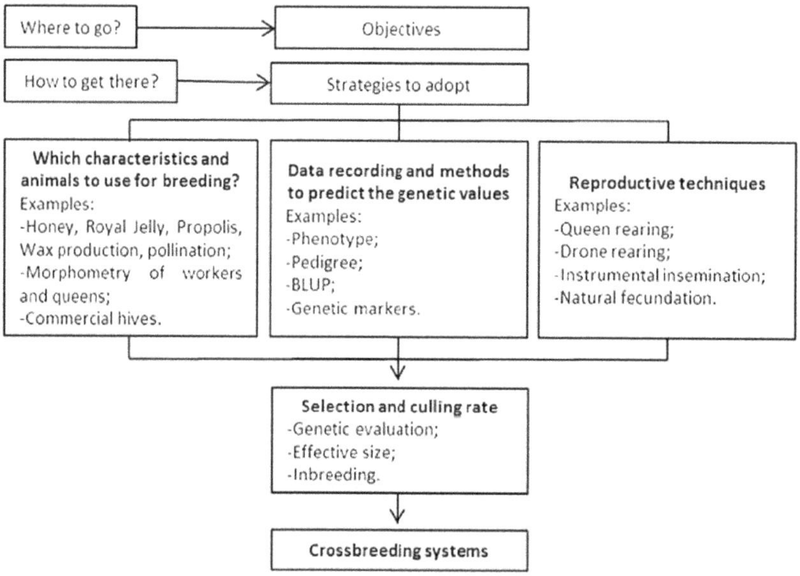

Figure 1. Strategies for honeybee breeding programs. Source: Ref. [8] adapted from Ref. [5].

There is a growing interest in starting a breeding program with Africanized honeybees in the country. Additionally, there is a need to find suitable and precise methods for calculating genetic factors that are of economic importance. Publications by Costa-Maia et al. [9], Faquinello et al. [10], Wielewski et al. [11], Garcia et al. [12], and Padilla et al. [13] are currently included in the body of research on estimating genetic parameters of Africanized honeybees in Brazil. These studies employ mixed models and Bayesian inference as their methodology.

Faquinello et al. [10] used Bayesian inference to estimate the genetic parameters of variance, covariance, heritability, and genetic correlations for royal jelly production in Apis mellifera L. Africanized. They found that royal jelly production is significantly influenced by the external environment and may also be affected by the internal environment, resulting in low heritability values for the analyzed characteristics. Furthermore, it is imperative for the beekeeper to exert control over these characteristics that need to be identified. Failure to do so will result in the inability to guarantee maximum production, even with the acquisition of queens that are suitable for royal jelly production. The researchers [10] discovered significant positive genetic correlations between colony-level royal jelly production and production per cup, with values of 0.29. They also observed a genetic connection of 0.42 between larvae acceptance and colony-level royal jelly production, and a low correlation of 0.06 between larvae acceptance and production per cup. The selection process, which focused on the feature of royal jelly output per colony, showed a favorable trend in genetic improvement for both larvae acceptance and royal jelly production per cup.

Conducting research on genetic parameters within a certain population enables the establishment of a selection procedure based on economically desirable traits. Consequently, there is a rise in the occurrence of favorable genes at the locus that holds economic significance. Consequently, the improvement in genetics is a result of the more accurate use of information about individuals throughout the selection process. This is due to the rising advancement in our understanding of genetic evaluation methods and the application of molecular genetics through the use of molecular markers.

Molecular genetics is crucial in enhancing the accuracy of information by identifying and investigating genes that have a significant impact on the manifestation of quantitative features. Lino-Lourenço and Costa-Maia [8] stated that the primary objective of genomic analysis in domestic animal species is to understand the genetic makeup of economically important traits. This involves determining the number of genes involved and their individual contributions to the expression of the phenotype. It also helps in recognizing that, in addition to numerous genes with small effects, there are a few genes that have a significant impact.

The use of phenotype as a basis for selection was enhanced by including information about genetic value, and subsequently, the integration of molecular genetics further enhanced the quality of information in animal breeding operations. In the future, we anticipate a rise in the gain per selection due to the ability to directly acquire the genetic merit of an animal from its genome in a comprehensive genetic evaluation procedure that includes all extant polymorphisms. Therefore, research on the production of royal jelly in the field of quantitative genetics, combined with molecular genetics, is highly significant for the development of breeding programs for Apis mellifera L. This study line necessitates not just laboratory work, but also a deep understanding of production management and the quality of royal jelly.

Royal jelly is a complex substance produced by the cephalic glands (hypopharyngeal and mandibular glands) of nurse honeybees (Apis mellifera L.). It is used to nourish the larvae during the first three days of their lives, while the queen bee is continuously fed with royal jelly during her whole lifespan. Royal jelly is a yellowish substance with a thick and creamy texture. It has a distinct smell and a slightly peppery taste [14]. The secretion is stimulated by the consumption of pollen together with regurgitated solutions from the nursing workers, which mostly include carbohydrates [15].

According to Fert (16), royal jelly has the ability to rejuvenate cells and tissues, regulate blood pressure, and enhance the generation of red blood cells. Royal jelly possesses a multitude of useful characteristics and has found extensive application in the fields of medicine, cosmetics, and nutritious cuisine across several nations [17]. Typically, it is composed of 60% water, proteins (41-42% when moisture is removed), carbs (30% when moisture is removed), and minor quantities of minerals, polyphenols, and vitamins [18]. Royal jelly plays a crucial role in the growth of queen honeybees by enhancing their fertility, lifespan, and physical size [19].

In order to produce royal jelly, honeybees require additional protein, particularly during periods of natural food scarcity. This protein can be sourced from pollen, as well as dairy products or brewer's yeast. Artificial cells are used to stimulate the production of royal jelly for queen rearing. These cells contain newly hatched larvae, which prompts worker honeybees to deposit royal jelly to nourish the growing larvae [20].

At the start of the twentieth century, Italian honeybees were brought to China. Chinese beekeepers highly value royal jelly, which is a significant product for them. They produce between 200 and 3500 tons of royal jelly every year, accounting for 90% of the world's total production. At first, the yield of royal jelly in China was a mere 0.2–0.3 kg per colony per year. The beekeepers in Zhejiang province initiated the process of choosing honeybees, and during a span of 20 years, the honey production per colony per year rose to 2.0–3.0 kg. Starting in 1980, the Chinese government recognized the significance of this product to the country and began investing in honeybee selection specifically for the production of royal jelly. This investment was made in the same province and other regions of the country. By the 2000s, production had reached a range of 6.0 to 8.0 kg per colony per year. Currently, the production has surpassed 10 kg per colony per year [23]. In order to achieve exceptional production values, Chen et al. [24] devised a production system consisting of eight steps. These steps encompass dietary supplementation, equipment modification or adjustment, skillful manipulation, and maintaining a high number of cups per colony.

Major Royal Jelly Proteins (MRJPs) constitute more than 90% of the total soluble protein content in royal jelly [25]. The identification of the genes responsible for encoding crucial proteins found in royal jelly was initiated through investigations conducted by Klaudiny et al. [26] and Albert et al. [27]. Following these groundbreaking results, other investigations are focused on characterizing novel genes that encode the MRJPs [25, 28, 29]. The locus Mrjp is a component of a cluster of nine genes that are encoded on a 65 kb array. It seems that the MRJP family has originated from a solitary ancestral gene that encodes a member of the yellow protein. There are five genes that encode proteins belonging to the yellow family. These genes are found in a specific area of the genome that also contains the genes responsible for generating MRJPs [30].

The five primary members of this family are MRJP1, MRJP2, MRJP3, MRJP4, and MRJP5 [25, 31]. While it has been suggested that royal jelly contains molecules that stimulate the queen's differentiation, there is less understanding of the role of its components, particularly the protein segment [32]. The MRJPs were identified and described in the species Apis mellifera L. [29, 30]. Nevertheless, Baitala et al. [33] have noted that there is a limited amount of data in the existing literature about the utilization of MRJPs as molecular markers in population genetic research and as selection markers for enhancing royal jelly output.

1.1. Factors that affect the creation of royal jelly

The formation of royal jelly is influenced by various elements, including internal parameters inside the colony, such as the posture and acceptance of larvae, as well as external ones including nutrition, climate conditions, temperature, precipitation, humidity, and even genetic factors [10, 34]. Toledo and Mouro [35] conducted an observation of identical environmental factors and found that higher external temperatures and relative humidity had a beneficial impact on royal jelly production, whereas rainfall had a negative effect.

Toledo et al. [34] found that the generation of royal jelly was not influenced by precipitation or the inclusion of a protein supplement (35%) when studying the impact of environmental variables. The number of accepted larvae was negatively impacted by the minimum relative humidity and maximum temperature, whereas the amount of accepted larvae was positively influenced by the maximum relative humidity [34]. Garcia and Nogueira-Couto [36] found that the acceptability of larvae for royal jelly manufacturing varies depending on the time of year.

The production of royal jelly is affected by several factors, including the frequency of collections per season, the geographical location of the apiary, the expertise of the beekeeper, and the genetic lineage of the honeybees [37]. Albarracín et al. [38] observed that larvae acceptance did not differ significantly between genetic groups of Italian and Africanized heritage. Additionally, the average production of royal jelly per colony was determined to be equivalent for both groups.

The money obtained from the time collection 48 hours after the larvae grafting was larger than the income obtained from the time collection 72 hours after the grafting, during a period of one month. Van Toor [20] proposed a method including two rounds of collection. The first collection should take place 72 hours after grafting the larvae, and the second collection should occur 48 hours after the first one. Any colonies that produce 180 mg or less of royal jelly per cup should be replaced. Nevertheless, Zheng et al. [21] proposed that the revised criteria for evaluating the quality of royal jelly in China should take into account the time of collection. This is because they observed notable variations in sample quality after 24, 48, and 72 hours of larvae grafting. Sereia et al. [39] found that the average quantity of royal jelly deposited every 68 hours per cup increased when supplemented with isolated soy protein, brewer's yeast, palm oil, and linseed oil. Several investigations have documented a range of royal jelly production in Africanized honeybees (Apis mellifera L.) from 188 to 234 mg per cup, as reported in references 10, 35, and 40. Toledo and Mouro [35] reported that the quantity of royal jelly collected each cup is influenced by the duration it remains within the colony. Africanized honeybees yielded an average of 253 mg/cup, while Carniolian honeybees produced an average of 198 mg/cup. Sereia et al. [39] found that grafted larvae showed varying levels of acceptance when exposed to a mixture of linseed oil + palm oil and isolated soy protein + brewer's yeast. The mixture of linseed oil + palm oil had a total acceptance rate of 63.45%, while the mixture of isolated soy protein + brewer's yeast had a total acceptance rate of 63.75%. In comparison, the acceptance rates for palm, linseed, isolated soy protein, yeast, and control groups I and II were 45.80%, 49.71%, 50.32%, 50.95%, 49.60%, and 52.17%, respectively.

In their study, Sereia et al. (2010a) [41] examined the impact of various nutrient supplements on the diet of honeybees. They discovered that the production of royal jelly, which is derived from glands, is influenced by the nutritional quality of the available sources. The researchers recommended protein supplementation to enhance royal jelly production in honeybee colonies. They also concluded that this practice is economically feasible for beekeepers [42]. Moreover, the use of dietary supplements in the production of royal jelly is a crucial aspect of the quality standards for this product in France [37]. In their study, Sereia et al. [41] assessed the nutritional value of supplements containing six distinct oil and protein sources. They found that a combination of linseed and palm oils, along with brewer's yeast, extended the lifespan of Africanized honeybee workers. Additionally, they observed that the use of sunflower oil increased the production of royal jelly [43]. The production of royal jelly is influenced by the biological and behavioral interactions of honeybees. It is also affected by genetic and environmental factors both within and beyond the colonies [34, 36].

1.2 Timeline of royal jelly manufacturing selection

Beekeeping, an emerging agribusiness in Brazil [34], holds significant potential for marketing various goods, including royal jelly, alongside honey. However, it is worth noting that royal jelly is not traditionally regarded a product of conventional beekeeping [42]. In addition to these items, Brazilian beekeepers are expanding their production of propolis and pollen due to the increasing pricing of these products.

In Brazil, a poly-hybrid has emerged from the crossing of four subspecies of European honeybees (Apis mellifera mellifera; A. m. caucasica; A. m. carnica; and A. m. ligustica) with African honeybees (Apis mellifera scutellata) [44]. This hybrid has undergone natural selection processes and adapted, resulting in what is now known as Apis mellifera L. Africanized honeybees. These bees have a high potential for selection due to their genetic diversity [35]. Garcia and Nogueira-Couto [36] found that African honeybees have successfully adapted to the environmental conditions in Brazil due to the similarities between Brazil and their native habitat. They have a high capacity for adaptation and pass this trait on to their offspring through crossbreeding with other subspecies. This has led to the widespread distribution of Africanized honeybees throughout the Americas. Africanized honeybees exhibit distinct variations compared to European honeybees, including variances in nest size and form, a shorter life cycle, colony growth and reproductive patterns, production of males, swarming behavior, defensiveness, and honey production in tropical climates.

Since 2006, Maringa State University has intensified studies on the evaluation system of Africanized honeybee colonies, which began in 1996. These studies involve the combined use of quantitative genetics and molecular genetics. The main objective is to secure funding for a breeding program that aims to selectively breed honeybees with economically desirable traits, such as increased royal jelly production. The Beekeeping Division of the Experimental Farm of Iguatemi at the University has begun to effectively compile data for the selection of Africanized honeybees (Apis mellifera L.) with the support of the Laboratory of Genetics and Cellular Biology.

The hybrid Carniolan honeybee colonies exhibited more royal jelly production compared to the Africanized honeybee colonies, as reported by Mouro and Toledo [40]. Nevertheless, there was a remarkable 109.19% surge in royal jelly output in the initial generation following the implementation of selection in Africanized honeybees. Hence, it is crucial to carefully choose the queen bees, and it is advised that the selection criteria should prioritize production per colony to prevent any loss in adaptive features such as disease resistance [40].

A study found that Africanized honeybees were more efficient in producing royal jelly compared to Carniolian honeybees, when specifically bred for royal jelly or honey production [35]. Furthermore, as a means of enhancing productivity, Toledo and Mouro [35] proposed the adoption of Africanized colonies due to the genetic variability found in the wild swarms or natural colonies of these honeybees in Brazil. Their research yielded significant outcomes that can be effectively implemented in practical settings. Toledo and Mouro [35] utilized Africanized honeybee colonies obtained from the wild to produce royal jelly between August 1996 and March 1998. Following the initial findings, the researchers picked the top five colonies that were most productive. These colonies were then compared to five Carniolan hybrid colonies, which included queen bees that originated from Germany. The researchers discovered that the Africanized honeybees generated a greater amount of royal jelly compared to the Carniolan honeybees.

Toledo et al. [43] conducted a study from August 2002 to February 2003 to assess the production of royal jelly in honeybees. They found that honeybees who were given diet supplemented with isoproteic (30%) and isolipid (5%) ingredients, specifically sunflower oil, showed a 28.79% increase in royal jelly output per colony. According to van Toor (20), it is recommended to choose colonies that have above-average production of royal jelly, taking into account factors such as colony size, high production of nurse honeybees, and favorable genetic predisposition.

Bayesian inference is a powerful tool that enhances the effectiveness of selection programs and has been employed in the assessment of animals to achieve more precise estimations. The data distribution enables the investigation of datasets of different sizes, allowing for precise estimation of variance components, breeding values, and confidence intervals [10, 46]. There is a growing interest in developing breeding programs for Africanized honeybees in the country. Additionally, there is a need to find suitable and precise methodologies for assessing the genetic parameters related to economic factors.

Faquinello et al. [10] employed Bayesian inference to estimate the variance components, covariance, and genetic factors for the generation of royal jelly in

Africanized honeybees using the animal model. During the experimental period, the authors [10] assessed many characteristics of royal jelly production in colonies with daughter queens from matrix colonies. The criteria measured were the acceptance of larvae, the production of royal jelly per colony, and the production of royal jelly per cup in both overlapped nucs (with five frames each) and overlapped Langstroth hives (with ten frames each). The selection process for queens was guided by genetic evaluation, with a focus on specific metrics such as larvae acceptance, royal jelly production by cup, and royal jelly production per colony [10]. Having a high-quality queen bee also leads to increased egg production, which in turn enhances the strength of the colony and maximizes overall productivity [47]. Sereia and Toledo (48) observed a genetic influence on royal jelly production. They determined that offspring from the most productive royal jelly colonies had a greater number of accepted larvae and deposited a bigger quantity of royal jelly in the cup.

The quantity of nurse honeybees in the colony has a direct impact on the production of royal jelly. Multiple surveys have reported elevated levels of mrjp expression or greater quantities of royal jelly protein in the hypo pharyngeal glands of nurse honeybees, a trait that is not observed in forager honeybees [49].

Baitala et al. [33] utilized microsatellite markers to detect seven alleles of mrjp3 in Africanized honeybees that produce royal jelly. They also verified the higher occurrence of alleles C, D, and E of Mrjp3 in specific colonies. The production results obtained from the genetic evaluations revealed that the examined queens exhibited comparable levels of royal jelly production. This suggests that there was no discernible variation in the alleles that were being selected for this particular trait. However, it was observed that the genotypes chosen for the matrix colonies, which included both queens and drones, were being passed on to the daughter queens. Parpinelli et al. [50] examined the genetic variability of the Mrjp3, Mrjp5, and Mrjp8 loci in colonies of Apis mellifera L. Africanized bees that have been selectively bred for royal jelly production since 2006. Their goal was to identify molecular markers that are associated with royal jelly production. They observed that these specific genetic traits became more prevalent during the process of selective breeding. Three sites exhibited polymorphism and yielded a combined total of 16 alleles. Additionally, four novel alleles have been discovered for the Mrjp5 locus [50]. The locus Mrjp3 has an effective number of alleles of 3.81. The mean observed heterozygosity was 0.4905, suggesting a substantial level of genetic diversity for the examined locus. The high values of the inbreeding coefficient (Fis) for the loci Mrjp3, Mrjp5, and Mrjp8 show an abundance of homozygotes. This suggests that the selection of Apis mellifera Africanized queens for royal jelly production is favoring the alleles mrjp3 C, D, and E, despite the fact that the C allele has occurred with a low frequency. Nevertheless, there is a decline in genetic diversity among the queens specifically at the studied locus, resulting in an overabundance of homozygotes. However, the high number of drones available for fertilizing the queens makes it difficult to produce homozygous genotypes for the Mrjp3 locus [50].

The research conducted at Maringa State University to optimize royal jelly production involves a multi-year assessment of honeybee colonies that are naturally fertilized. This assessment includes quantitative and molecular analysis, with a focus on genetic markers for protein MRJP3. These markers are used to select colonies that have the highest quantitative production of royal jelly.

Researchers, including Mouro and Toledo [40], Toledo and Mouro [35], Baitala et al. [33], Toledo et al. [34], Faquinello et al. [10], Toledo et al. [43], and Parpinelli et al. [50], have contributed to the ongoing study in this field. The honeybee selection process involves controlled crosses utilizing instrumental insemination procedures to obtain homozygous individuals. This allows us to identify the allele that contributes the most to the significant increase in production. However, it is crucial to occasionally introduce a queen with distinct genetic traits in order to maintain heterosis.

This study aimed to forecast the genetic worth of Africanized Apis mellifera L. honeybee colonies that produce royal jelly. The prediction was based on genetic information obtained from production characteristics such as larvae acceptance and royal jelly production per colony and per cup. The data used for this analysis were collected between 2006 and 2011.

2. MATERIAL AND METHODS

The experiment took place at the Experimental Farm of Iguatemi at Maringa State University, Brazil, from January to April 2011. The data used to assess the Africanized honeybees and predict breeding values for royal jelly production were collected from 2006 to 2011. During this time, genomic DNA extractions were also performed on the selected colonies using molecular genetic markers.

2.1. Identification of genomic DNA

The Laboratory of Genetics and Cellular Biology at the University provided the genetic information of specific colonies that were chosen and assessed, as reported by Baitala et al. [33] and Parpinelli et al. [50]. These scientists utilized the technique outlined by Bardakci and Skibiński [51] to extract the whole DNA from the thorax of nursing honeybees. The method was modified to suit the species Apis mellifera L.

The polymerase chain reaction (PCR) was conducted using custom-designed primers to amplify the repetitive regions of the Mrjp3 locus [28]. The amplification reactions were carried out in a Techne thermal cycler TC-512, following the amplification conditions for primer MRJP3 as described by Albert and Schmitz [52].

2.2. Queen rearing and royal jelly production

Queen rearing was conducted in a dedicated facility to cultivate this activity in a controlled atmosphere with air conditioning. The larvae grafting process took place at an average temperature of 33 ± 2°C and a relative humidity of 60 ± 10%. Beekeepers from several regions including Parana, São Paulo, Mato Grosso, Mato Grosso do Sul, Sergipe, Paraguay, and Colombia contributed colonies. Each time a colony or swarm perished, it was subsequently substituted by another one. Each colony was recognized and the queens were labeled with numbered plates on their thorax. The colonies were housed in either a five frame hive or a Langstroth hive, depending on their size and the time of year. Subsequently, we commenced the process of raising the initial cohort of female offspring derived from those matriarchs. The technique employed for the creation of queens was modified from Doolittle's method [47]. The grafting procedure was uncomplicated, using larvae that were between 0 and 24 hours old and had controlled lineages. Regularly, the genealogy of all colonies was determined using genotyping.

The colonies that were established and completed were referred to as mini-hives [53]. For the purpose of queen raising, a total of 10 queens were produced from each selected colony in every generation. Each colony was established by combining two overlapping nucleus colonies with a queen excluder separating them. The mini-hive colony consisted of a cup bar frame with 30 acrylic cups. There were 15 cups in the upper bar and 15 cups in the lower bar, each with distinct genealogies. The cups were identified and distributed randomly, as shown in Figure 2. After a period of 10 days, the cells containing the queens were extracted from the colonies where they were being raised and transferred to incubators. The queens were allowed to develop inside glass vials of 20 mL, each accompanied by a piece of paper that provided information about their genealogy and hive number. Recently hatched queens were rendered unconscious using carbonic gas (CO2), labeled with a unique number on the top thorax, kept in a plastic enclosure, and stored until they were introduced into the colonies for the purpose of producing royal jelly.

Figure 2. Withdrawn of frame with queen cells for queen production.

Following the emergency, the queens were sedated using carbon dioxide (CO2) in order to measure their body weight (in milligrams) using a very accurate digital scale with a precision of 0.001 grams. Additionally, their length and abdominal width (in millimeters) were measured using a digital precision caliper with a precision of 0.01 millimeters or 0.0005 inches. Queens weighing more than 180 mg were placed in JZsBZs™ type cages and housed in an incubator with nurse honeybees. They were then transferred into five-frame hives for natural mating. This introduction was conducted at least 24 hours after the supersedure to decrease the potential of plunder and during a period of heightened acceptance [47]. Since the queens were fertilized through natural means, the details about the fathers' genealogy were regarded as unknown.

Monitoring of oviposition initiation was conducted to commence evaluations of royal jelly production after a period of 50 days, ensuring that all worker honeybees were offspring of the newly established queen. Terada et al. [54] confirmed that the mean lifespan of an Africanized worker is 26.3 days.

In order to produce royal jelly, a frame was put into each colony. This frame consisted of three bars and a total of 100 fake cups. After a period of 66-72 hours, the frames were taken out (as shown in Figure 3), the larvae were disposed of, and the royal jelly was collected using a suction apparatus. The larvae grafting procedure was planned according to the schedule outlined by Wielewski et al. [11].

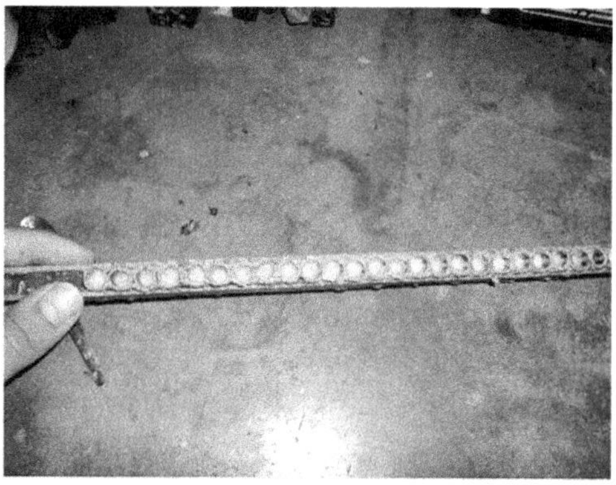

Figure 3. Bar with cups showing the larvae acceptance and royal jelly produced.

2.3. Statistical analysis

The weight, length, and width data of the abdomen in all generations of selected newly emerging queens were analyzed using analysis of variance (ANOVA) with SAS software [55]. The average values for each evaluated attribute in each generation were compared using Tukey's test at a significance level of 5%. Once the data on body weight, length, and abdomen width of the newly emerged queens were collected, a genetic evaluation was conducted using the software Multiple Trait Gibbs Sampling in Animal Models (MTGSAM). This software, developed by

Van Tassel & Van Vleck [56], utilizes the Bayesian estimation approach through the Gibbs sampling method.

The animal model used was in the following:

$$Y = X\beta + Za + e$$

where *y* is the vector of observations; *X* is the incidence matrix of fixed effects, contained in the vector β; β is the vector of fixed effects; *Z* is an incidence matrix of additive genetic effects; *a* is the vector of additive genetic effects; and *e* is the vector of random errors associated to each observation.

where *y*, *a*, and *e* have normal multivariate joint distribution, as follows:

$$\begin{bmatrix} y \\ a \\ e \end{bmatrix} \sim NMV\left\{\begin{bmatrix} X\beta \\ 0 \\ 0 \end{bmatrix};\begin{bmatrix} ZGZ'+R & ZG & R \\ GZ' & G & 0 \\ R & 0 & R \end{bmatrix}\right\}$$

In unicaracter analysis, *G* is the genetic variance and covariance matrix as $A\sigma^2_a$, *A* being relationship matrix, and σ^2_a is the additive genetic variance; *R* is the residual variance matrix given by $I\sigma^2_a$, *I* being identity matrix, and σ^2_a is the residual variance of the trait.

For bicaracter analysis, the *G* matrix is $G_0 \otimes A$, *A* being the relationship matrix, and G_0 is the additive genetic covariance matrix as follows:

$$G_0 = \begin{bmatrix} \sigma^2_{a_1} & \sigma_{a_1a_2} \\ \sigma_{a_2a_1} & \sigma^2_{a_1} \end{bmatrix}$$

The matrix R is given by $R_0 \otimes I$, *I* being identity matrix by equal order to drone number, and R_0 is the residual covariance matrix, as below:

$$R_0 = \begin{bmatrix} \sigma^2_{e_1} & \sigma_{e_1e_2} \\ \sigma_{e_2e_1} & \sigma^2_{e_1} \end{bmatrix}$$

In analysis strategy, it was used Gibbs chains of 58,500 cycles resulting from 650,000 cycles were generated, respectively, with initial disposal of 65,000 iterations and sampling intervals for every 10 iterations. The convergence of chains was tested by Heidelberger and Welch [57] test, implemented in Convergence Diagnosis and Output Analysis (CODA)—R software—Version 2.12.1. [58].

The colonies were categorized into upper and lower groups based on expected values, which were determined by averaging the genetic values of the parameter used as the classification criteria. The genotype's probability rating in the higher and lower classes was evaluated using the PROC GENMOD routine from the Statistical Analysis System [55]. The data was assumed to follow a binomial distribution with logarithmic linkage. The classification probabilities for each genotype were assessed using a T test. This was done using the PROC GENMOD routine with a significance level of 5%.

The genetic data were compared across different genotypes using PROC GENMOD procedure, and the resulting averages were calculated by several compilation tests. In order to minimize the impact of environmental factors on royal jelly production, a predictive model was employed to estimate breeding values. The model took into account the environmental effects of year (2006-2011), time (the four seasons of the year), and type (colony model) with a flat distribution. Additionally, the collection and distribution of chi square were considered.

The MTGSAM software utilized Bayesian inference to determine the genealogy and forecast breeding values for each colony in several features, such as larvae acceptance per grafting (%), royal jelly production per colony (g), and royal jelly production per cup (mg). Utilizing this form of analysis is suitable for enhancing the precision of the dataset that adheres to this particular beekeeping analysis technique.

The SAS program was utilized to ascertain the disparities in production metrics in relation to the anticipated values for the assessed colonies. Given that each colony functions as a superorganism [59], the study involved analyzing a total of 78 unique units over a period of 6 years, from 2006 to 2011. The most commonly seen genotypes in the colonies selected for royal jelly production over the years were DE, DC, CE, EF, and FG [50].

3. RESULTS

3.1. Performance rating

Table 1 displays the statistical disparities seen in the production characteristics of the chosen colonies, comparing the high and low genetic value classes, while taking into account the genotypes. There were no significant statistical differences in the percentage of total larvae acceptance per grafting among the DE and EF genotypes. However, the FG genotype performed the worst in this parameter, with a higher likelihood of falling below the average. Regarding this situation, it is noteworthy that the DC genotype exhibits a 71% probability of being ranked higher than the average of the acceptance values. The total royal jelly output per colony (g) showed consistent performance across different genotypes, with no statistically significant differences. However, the DE genotype stood out with a 48% chance of being categorized as having high genetic value for this metric. When comparing the royal jelly production per cup, the DE genotype demonstrated the highest rating, with a 60% probability of being over the average. This result was considerably different from the EF genotype and did not show any significant difference from the other genotypes. The EF genotype exhibited the lowest rating, with a 78% probability of being categorized as having low genetic value for this metric.

Table 1. Probability of classification of high and low genetic value of different genotypes for total larvae acceptance per grafting, royal jelly production per colony, and royal jelly production per cup.

Genotype	Class	Evaluated parameters		
		Total larvae acceptance per grafting (%)	Royal jelly per colony (g)	Royal jelly per cup (mg)
DE	High-genetic value	0.50a	0.48a	0.60a
	Low-genetic value	0.50	0.52	0.40
DC	High-genetic value	0.71a	0.43a	0.43ab
	Low-genetic value	0.29	0.57	0.57
CE	High-genetic value	0.50a	0.25a	0.25ab
	Low-genetic value	0.50	0.75	0.75
EF	High-genetic value	0.44a	0.33a	0.22b
	Low-genetic value	0.56	0.67	0.78
FG	High-genetic value	0b	0b	0.33ab
	Low-genetic value	1	1	0.67

Means followed by the same letters within the classes, in the same column, are not statistically different from each other by *T* test ($P > 0.05$).

3.2. Prediction of genetic values

The table displays the mean values of expected genetic values for each genotype in Table 2. The DE and DC genotypes exhibited better genetic values compared to the EC, EF, and FG genotypes, thereby leading to larvae acceptance. Genotype DE was indistinguishable from all other genotypes. The DC genotype exhibited a substantial difference compared to the FG genotype, which demonstrated the poorest performance in this measure. The DE genotype exhibited greater total royal jelly production per colony, distinguishing it from the DC and CE genotypes, but showing no significant difference compared to the EF and FG genotypes.

Table 2. Average genetic values for the characteristics evaluated in terms of different genotypes.

Genotype	Genetic values		
	Total larvae acceptance per grafting (%)	Royal jelly per colony (g)	Royal jelly per cup (mg)
DE	0.8043ab	0.4505a	0.0234a
DC	2.3118a	-0.0955ab	-0.007b
CE	-6.7991ab	-1.1013b	-0.0388b
EF	-4.2494ab	-0.612b	-0.0109ab
FG	-9.2538b	-1.2472b	-0.0283ab

Means followed by the same letter, in the same column, do not differ statistically from each other by *T* test ($P > 0.05$).

4. DISCUSSION

Bayesian inference is becoming more popular in honeybee husbandry trials because it provides precise estimates of variance components, genetic values, and credibility intervals. This is particularly useful when working with a small sample size and considering data distribution, as it contributes to an efficient selection program. Nevertheless, until recently, there have been no studies undertaken on the generation of royal jelly in honeybees using these approaches [10, 11]. Metorima et al. [61] suggest using this method for data that does not have this type of limitation. It should be employed as a tool to obtain more precise estimations for honeybee research.

Genetic value refers to the portion of the genotypic value that is passed down from parents to their offspring. Genetic value prediction facilitates comparisons between statistical analyses of quantitative genetics and molecular genetics. According to Costa-Maia et al. [9], precise measurement of genetic parameters enables the prediction of an animal's genetic worth, which in turn allows for the identification of individuals with superior genetic traits.

Laboratory investigations have shown that specific alleles are becoming less common while others are becoming more common, as seen by an increase in the frequency of these relevant alleles in royal jelly production. Baitala et al. [33] and Parpinelli et al. [50] observed that the frequency of alleles C, D, and E increased in nurse honeybees after DNA extraction. These bees were included in a system for evaluating royal jelly output based on both physical characteristics and genetic analysis.

The decrease in heterozygosity is responsible for the increased occurrence of alleles C, D, and E, while simultaneously decreasing the frequency of alleles F and G. By examining the royal jelly output per colony, it is evident that the D allele plays a significant role, along with the DE and DC genotypes which exhibited notable contributions. During the genetic evaluations for royal jelly production, alleles that were most closely associated with higher output were consistently retained throughout the selection process, as stated by Parpinelli et al. [50]. Parpinelli et al. [50] found that selecting for royal jelly production in MRJPs is causing an increase in homozygosity at these loci, particularly MRJP3, which has the lowest observed heterozygosity.

When evaluating the acceptance of larvae after grafting, only the FG genotype was considered poor, while all others were categorized as high. The DC genotype stood out as superior, with a 71% chance of being classified as such. A comparative survey conducted by Baumgratz et al. [62] examined the Doolittle and "Starter" methods for royal jelly production. The study indicated that the first method had a larvae acceptance percentage of 63%, while the second approach had a percentage of 50%. These figures were higher than the 29.20% recorded by Toledo et al. [34].

Faquinello et al. [10] found that the average larvae acceptance rate was 52.13%, which was consistent with the performance of nearly all genotypes examined in our investigation. Sereia et al. [39] found that the acceptance of grafted larvae differed when exposed to a mixture of linseed oil + palm oil and isolated soy protein + brewer's yeast. The acceptance rates for these mixtures were 63.45% and 63.75%, respectively, compared to acceptance rates of 45.80%, 49.71%, 50.32%, 50.95%, 49.60%, and 52.17% for palm, linseed, isolated soy protein, yeasts, and controls I and II.

For Faquinello et al. [10], the average royal jelly production per colony was 6.26 g and the average royal jelly production per cup was 190.07 mg, both falling under the same classification as previously mentioned. This was because the DE genotype showed a statistically significant difference solely from the EF genotype, with no difference observed between the other genotypes. Baumgratz et al. [62] recorded yields of 268 and 269 mg per cup. Faquinello et al. [10] discovered that there were positive correlations between the amount of royal jelly produced per colony and the amount of royal jelly produced per cup. They also found a correlation between the rate of acceptance of larvae after grafting and the amount of royal jelly produced per colony. However, there was a low correlation between the rate of acceptance of larvae after grafting and the amount of royal jelly produced per cup. In their study, Muli et al. [63] compared the royal jelly production potential of two subspecies of African Apis mellifera L. They discovered that collecting the royal jelly after 3 days resulted in a larger yield per cup, with 349.5 mg, compared to collecting after 2 days, which yielded 236.3 mg.

The production of royal jelly is significantly impacted by the surrounding environment. According to the genetic correlation, the selection process resulted in an increase in royal jelly production per colony, larvae acceptance, and royal jelly production per cup [10]. According to Li et al. [64], the hypo pharyngeal glands of nurse honeybees chosen for royal jelly production were substantially larger than those of non-selected honeybees. Microscope pictures revealed that the royal jelly secretion period was greater in selected honeybees compared to non-selected honeybees [65]. This emphasizes the significance of carefully choosing and evaluating individuals based on their genetic traits, as well as initiating a breeding program specifically for honeybees, particularly those of Africanized descent.

The average genetic values obtained for the three criteria were positive for the DE genotype, followed by the DC genotype, which also showed a positive average value for larvae acceptance, indicating higher classification. All other genotypes exhibited negative results for the average genetic value, indicating that they were evaluated lower. Regarding the overall acceptance of larvae following the grafting process, the only notable variations were observed between the DC and FG genotypes, with the DC genotype exhibiting better acceptability compared to the other genotypes. To boost royal jelly output per colony when there is little acceptance of larvae, one might increase the amount of royal jelly deposited per cup. The genetic values for royal jelly production were predicted, and the use of genetic evaluation tools revealed that the alleles D and E, specifically related to MRJP, have the highest genetic value for producing royal jelly.

The interplay between environmental factors and genetic variations poses challenges to honeybee reproduction, specifically in terms of mating levels [67]. Royal jelly production is a heritable characteristic that may be controlled genetically [68]. Due to its significant economic worth, there is a demand for tools to develop breeding programs [4, 33, 34, 40, 42]. According to Harbo and Rinderer (69), the process of choosing better honeybee genotypes entails utilizing advanced methods such as enhanced queen replacement techniques, artificial insemination, and assisted selection with molecular markers. However, there is a scarcity of evidence in the literature about molecular markers that are associated with royal jelly production [4, 10, 11, 33, 34, 40, 42, 70]. The choice of more efficient queens is advantageous. In order to engage in beekeeping and effectively manage breeding control and selection, it is essential to assess numerous colonies across multiple apiaries, maintain meticulous records, and ideally have access to an insemination laboratory.

Several studies, including Mouro and Toledo [40], Toledo and Mouro [35], Baitala et al. [33], Toledo et al. [34], Faquinello et al. [10], Toledo et al. [43], Parpinelli et al. [50], and the current research with predicted values, are significant and should be considered for the implementation of breeding programs. The availability of this data enables the research to proceed by selecting colonies based on the forecast of their genetic values. Nevertheless, by employing controlled crossings through artificial insemination procedures, it becomes feasible to acquire individuals that are homozygous for the allele responsible for the most significant enhancement in royal jelly output. In addition, the significant variation in alleles for the MRJP3 protein suggests that this biomolecular marker can be utilized in investigations of the genetic composition of Africanized honeybees. Consequently, following the selection process, it is recommended to implement a breeding program [33, 70] based on this marker. It is advisable for the government to endorse this technology to ensure its widespread adoption among beekeepers in the country.

Based on the analyses, it is shown that alleles D, E, and C have the greatest significance in relation to the evaluated production characteristics, specifically larvae acceptance after grafting and royal jelly production per colony and per cup. Therefore, it is recommended to retain the DE, DC, and EC genotypes in the royal jelly production evaluation system, whereas the remaining genotypes should be eliminated or substituted due to their poor performance in these crucial production characteristics.

REFERENCES

1. Meixner MD, Buchler R, Costa C, Francis RM, Hatjina F, Kriger P, Uzunov A, Carreck NL. Honey bee genotypes and the environment. Journal of Apicultural Research 2014;53:183–187. DOI: 10.3896/IBRA.1.53.2.01
2. Almeida MAD, Carvalho CMS. [Beekeeping: a sustainable business opportunity, in Portuguese]. Apicultura: uma oportunidade de negócio sustentável. Salvador: SEBRAE Bahia; 2009. 50 p.
3. Rinderer TE. Bee genetics and breeding. New York: Academic Press; 2008. 426 p.
4. Widdicombe J. The principles of bee improvement. Mytholmroyd: Northern Bee Books, 2015. 80 p.
5. Kinghorn B, Werf J, Ryan M. [Animal Breeding - use of new technologies, in Portuguese]. Melhoramento animal – uso de novas tecnologias. Piracicaba: FEALQ; 2006. 367 p.
6. Snelgrove LE. Queen rearing. Mytholmroyd: Lightning Source UK and Northern Bee Books; 2008. 344p.
7. Mahbobi A, Farshineh-Adl M, Woyke J, Abbasi S. Effects of the age of grafted larvae and the effects of supplemental feeding on some morphological characteristics of Iranian queen honey bees (Apis mellifera meda Skorikov, 1929). Journal of Apicultural Science 2012;56:93–98. DOI: 10.2478/v10289-012-0010-1
8. Lino-Lourenço, DA; FM, Costa-Maia, FM. [Genomics: the application of genetic advances in animal breeding, in Portuguese]. Genômica: a aplicação dos avanços da genética no melhoramento animal. In: Martin TN, Ziech MF, Pavinato OS, Waclawovsky AJ, Sklarski MMSS, org. [Agricultural Production Systems, in Portuguese]. Sistemas de Produção Agropecuária 2009. Curitiba: UTFPR; 2009. p. 90-112.
9. Costa-Maia FM, Toledo VAA, Martins EN, Lino-Lourenço DA, Sereia MJ, Oliveira CAL, Faquinello P, Halak AL. Estimates of covariance components for hygienic behavior in Africanized honeybees (Apis mellifera). Revista Brasileira de Zootecnia 2011;40:1909–1916. DOI: 10.1590/S1516-35982011000900010
10. Faquinello P, Toledo VAA, Martins EN, Oliveira CAL, Sereia MJ, Costa-Maia FM, Ruvolo-Takasusuki MCC. Parameters for royal jelly production in Africanized honeybees. Sociobiology 2011;57:495–509.
11. Wielewski P, Toledo VAA, Martins EN, Costa-Maia FM, Faquinello P, Lino-Lourenço DA, Ruvolo-Takasusuki MCC, Oliveira CAL, Sereia MJ. Relationship between hygienic behavior and Varroa destructor mites in colonies producing honey or royal jelly. Sociobiology 2012;59:251–274.

12. Garcia RC, Oliveira NTE, Camargo SC, Pires BG, Oliveira CAL, Teixeira RA, Pickler MA. Honey and propolis production, hygiene and defense behaviors of two generations of Africanized honey bees. Scientia Agricola 2013;70:74–81. DOI: 10.1590/S0103-90162013000200003

13. Padilha AH, Sattler A, Cobuci JA, Mcmanus CM. Genetic parameters for five traits in Africanized honeybees using Bayesian Inference. Genetics and Molecular Biology 2013;36:207–213. DOI: 10.1590/s1415-47572013005000016

14. Mărghitaş LAl. [Bees and their products, in Romanian]. Albinele şi produsele lor. Bucureşti: Ceres; 2008. 391 p.

15. Haydak MH. Honey bee nutrition. Annual Review of Entomology 1970;15:143–156. DOI: 10.1146/annurev.en.15.010170.001043

16. Fert G. Cría de reinas. Buenos Aires: Campo and Abejas; 2011. 128 p.

17. Ramadan MF, Al-Ghamdi A. Bioactive compounds and health promoting properties of royal jelly: a review. Journal of Functional Foods 2012;4:39–52. DOI: 10.1016/j.jff.2011.12.007

18. Sabatini AG, Marcazzan GL, Carboni MF, Bogdanov S, Almeida-Muradian LB. Quality and standardization of royal jelly. Journal of Api Product and Api Medical Science 2009;1:16–21. DOI: 10.3896/IBRA.4.1.01.04

19. Muresan CL, Marguitas LAI, Dezmirean DS, Bobis O, Bonta V, Zacharias I, Margaoan R, Pasca C. Quality parameters for commercialized royal jelly. Bulletin UASVM Animal Science and Biotechnologies 2016;73:1–8. DOI: 10.15835/buasvmcn-asb:11630

20. van Toor RF. Producing royal jelly. Tauranga: Bassdrum Books; 2006. 103 p.

21. Zheng HQ, Hu FL, Dietmann V. Changes in composition of royal jelly harvested at different times: consequences for quality standards. Apidologie 2011;42:39–47. DOI: 10.1051/apido/2010033

22. Apicultural Science Association of China (ASAC). 2013. China's bee products market review of 2012 and prospect for 2013. Beijing: ASAC; 2013.

23. Chen MH. Analysis of the current state of royal jelly in China mainland. Journal of Bee 2005;25:17–19.

24. ChenS,SuS,LinX.Anintroductiontohigh-yieldingroyaljellyproductionmethods in China. Bee World 2002;83:69–77. DOI: 10.1080/0005772x.2002.11099543

25. SchmitzovàJ, Klaudiny J, Albert S, Schröder W, Schreckengost W, Hanes J, JúdovàJ, Simúth J. A family of major royal jelly proteins of the honeybee Apis mellifera L. Cellular and Molecular Life Sciences 1998;54:1020–1030. DOI: 10.1007/s000180050229

26. Klaudiny J, Hanes J, Kulifajova J, Albert S, Simúth J. Molecular cloning of two c DNAs from the head of the nurse honey bee (Apis mellifera L.) coding for related proteins of royal jelly. Journal of Apicultural Research 1994;33:105–111. DOI: 10.1080/00218839.1994.11100857

27. Albert S, Klaudiny J, Simúth J. Newly discovered features of the updated sequence of royal jelly protein RJP57-1, longer repetitive region on C-terminus and homology to Drosophila melanogaster yellow protein. Journal of Apicultural Research 1996;35:63–68. DOI: 10.1080/00218839.1996.11100914

28. Albert S, Klaudiny J, Simúth J. Molecular characterization of MRJP3, highly polymorphic protein of honeybee (Apis mellifera) royal jelly. Insect Biochemistry and Molecular Biology 1999;29:427–434. DOI: 10.1016/S0965-1748(99)00019-3

29. Albert S, Klaudiny J. The MRJP/yellow protein family of Apis mellifera: identification of new members in the EST library. Journal of Insect Physiology 2004;50:51–59. DOI:10.1016/j.jinsphys.2003.09.008

30. Drapeau MD, Albert S, Kucharski R, Prusko C, Maleszka R. Evolution of the yellow/major royal jelly protein family and the emergence of social behavior in honey bees. Genome Research 2006;16:1385–1394. DOI: 10.1101/gr.5012006

31. Buttstedt A, Moritz RFA, Erler S. Origin and function of the major royal jelly proteins of the honeybee (Apis mellifera) as members of the yellow gene family. Biological Reviews 2014; 89:255–269. DOI:10.1111/brv.12052

32. Blank S, Bantleon FI, Mc Intyre M, Ollert M, Spillner E. The major royal jelly proteins 8 and 9 (Api m 11) are glycosylated components of Apis mellifera venom with allergenic potential beyond carbohydrate-based reactivity. Clinical and Experimental Allergy 2012;42:976–985. DOI: 10.1111/J.1365-2222.2012.03966.x

33. Baitala TV, Faquinello P, Toledo VAA, Mangolin CA, Martins EN, Ruvolo-Takasusuki MCC. Potential use of major royal jelly proteins (MRJPs) as molecular markers for royal jelly production in Africanized honeybee colonies. Apidologie 2010;41:160–168. DOI: 10.1051/apido/2009069

34. Toledo VAA, Neves CA, Alves EM, Oliveira JR, Ruvolo-Takasusuki MCC, Faquinello P. [Royal jelly production in Africanized honeybee colonies considering different protein supplements and the influence of environmental factors, in Portuguese]. Produção de geleia real em colônias de abelhas africanizadas considerando diferentes suplementos proteicos e a influência de fatores ambientais. Acta Scientiarum. Animal Sciences 2010;32:93–100. DOI: 10.4025/actascianimsci.v32i1.6836

35. Toledo VAA, Mouro GF. [Royal jelly production by selected Africanized honeybees and Carniolan hybrids, in Portuguese]. Produção de geleia real com abelhas africanizadas selecionadas e cárnicas híbridas. Revista Brasileira de Zootecnia 2005;34:2085–2092. DOI:10.1590/S1516-35982005000600034

36. Garcia RC, Nogueira-Couto RH. [Royal jelly production by Apis mellifera Italian, Africanized and descendants of their crosses, in Portuguese]. Produção de geleia real por abelhas Apis mellifera italianas, africanizadas e descendentes de seus cruzamentos. Acta Scientiarum. Animal Sciences 2005;27:17–22. DOI: 10.4025/actascianimsci.v27i1.1254

37. OPIDA. [The guide techinique of royal jelly producer, in French]. Le guide techinique du producteur de gelée royale. Lion: Office Pour L'information et la Documentation en Apiculture; 2009. 117 p.

38. Albarracín VN, Funari SRC, Arouco EMR, Orsi RO. Acceptance of larvae from different genetic groups of Apis mellifera in queen bee production. Archivos Latinoamericanos de Producción Ani mal 2006;14:33–41.

39. Sereia MJ, Arnaut de Toledo VA, Furlan AC, Faquinello P, Costa-Maia FM, Wielewski P. Alternative sources of supplements in Africanized honeybees submitted to royal jelly production. Acta Scientiarum. Animal Sciences 2013;35:165–171. DOI: 10.4025/actascianimsci.v35i2.16976

40. Mouro GF, Toledo VAA. Evaluation of Apis mellifera Carniolan and Africanized honey bees in royal jelly production. Brazilian Archives of Biology and Technology 2004;47:469–476. DOI: 10.1590/s1516-89132000300018

41. Sereia MJ, Arnaut de Toledo VA, Faquinello P, Costa-Maia FM, Satilla ESC, Ruvolo-Takasusuki MCC, Furlan AC. Lifespan of Africanized honey bees fed with various proteic supplements. Journal of Apicultural Science 2010;54:37–49.

42. Sereia MJ, Arnaut de Toledo VA, Ruvolo-Takasusuki MCC, Sekine ES, Faquinello P, Costa-Maia FM. [Financial viability of the royal jelly production with Africanized honeybees supplemented with different nutrients, in Portuguese]. Viabilidade financeira da produção de geleia real com abelhas africanizadas suplementadas com diferentes nutrientes. Acta Scientiarum. Animal Sciences 2010;32:467–474. DOI: 10.4025/actascianimsci.v32i4.7895

43. Arnaut de Toledo VA, Mello AIP, Sales PJP, Costa-Maia FM, Ruvolo-Takasusuki MCC, Furlan AC, Faquinello P. [Royal jelly production in Africanized honeybees considering different polyunsaturated fatty acids and environmental factors, in Portuguese]. Produção de geleia real em abelhas africanizadas considerando diferentes ácidos graxos poliinsaturados e fatores ambientais. Global Science and Technology 2012;5:164–175.

44. Kerr WE. [Introduction of African honeybees in Brazil, in Portuguese]. Introdução de abelhas africanas no Brasil. Brasil Apicola 1957;3:211–213.

45. Sanford MT. The Africanized honey bee in the Americas: a biological revolution with human cultural implications. American Bee Journal 2006;146:209–211.

46. van Tassel CP, van Vleck LD. A manual for use of MTGSAM. A set of FORTRAN programs to apply Gibbs sampling to animal models for variance component estimation. Lincoln: Department of Agriculture/Agricultural Research Service; 1995. 86 p.

47. Doolittle GM. Scientific queen-rearing. Chicago: Thomas G. Newman and Son; 2008. 101 p. (reprint from 1889 and 1899)

48. Sereia MJ, Arnaut de Toledo VA. Quality of royal jelly produced by Africanized honeybees fed a supplemented diet. Ciência e Tecnologia de Alimentos (Online) 2013;33:304–309. DOI: 10.1590/S0101-20612013005000039

49. Ueno T, Nakaoka T, Takeuchi H, Kubo T. Differential gene expression in the hypopharyngeal glands of worker honeybees (Apis mellifera L.) associated with an age-dependent role change. Zoological Science 2009;26:557–563. DOI: 10.2108/zsj.26.557

50. Parpinelli RS, Ruvolo-Takasusuki MCC, Toledo VAA. MRJP microsatellite markers in Africanized Apis mellifera colonies selected on the basis of royal jelly production. Genetics and Molecular Research 2014;13:6724–6733. DOI: 10.4238/2014.August.28.16

51. Bardakci F, Skibinski DO. Application of the RAPD technique in tilapia fish: species and subspecies identification. Heredity 1994;73:117–123. DOI: 10.1038/hdy.1994.110

52. Albert S, Schmitz J. Characterization of major royal jelly protein-like DNA sequences in Apis dorsata. Journal of Apicultural Research 2002;41:75–85. DOI: 10.1080/00218839.2002.11101072

53. Laidlaw HH, Eckert JE. Queen rearing. Berkeley: University of California Press; 1962.

54. Terada Y, Garofalo CA, Sakagami SF. Age-survival curves for workers of two eusocial bees (Apis mellifera and Plebeia droryana) in a subtropical climate, with notes on worker polytheism in P. droryana. Journal of Apicultural Research 1975;14:161–170. DOI: 10.1080/00218839.1975.11099821

55. Statistical Analysis System – SAS Institute Inc. Base SAS® 9.4 procedures guide: statistical procedures. 2nd Ed. Cary, NC: SAS Institute Inc.; 2013.

56. Van Tassel CP, Van Vleck LD. A manual for use of MTGSAM. A set of Fortran programs to apply Gibbs sampling to animal models for variance component

estimation (DRAFT). Lincoln: Department of Agriculture/Agricultural Research Service; 1995, 86 pp.

57. Heidelberger P, Welch PD. Simulation run length control in the presence of an initial transient. Journal of the Operations Research Society of America 1983;31:1109–1144. DOI: 10.1287/opre.31.6.1109

58. R Development Core Team. R: a language and environment for statistical computing. Vienna, Austria: R Foundation for Statistical Computing; 2009. ISBN 3-900051-07-0. http://www.R-project.org

59. Moritz RFA, Southwick EE. Bees as superorganisms. New York: Springer-Verlag; 1992. 395 p. DOI: 10/1007/978-3-642-84666-3

60. Rekaya R, Weigel KA, Gianola D. Bayesian estimation of parameters of a structural model for genetic covariances between milk yields in five regions of the United States. Journal of Dairy Science 2003;86:1837–1844. DOI: 10.3168/jds.s0022-0302(03)73770-9

61. Metorima FN, Costa-Maia FM, Halak AL, Parpinelli RS, Toledo VAA. Morphometric measurements of Africanized honeybee queens kept in an incubator or in queen banking. Acta Scientiarum. Animal Sciences 2015;37:91–96. DOI: 10.4025/actascianimsci.v37i1.24006

62. Baumgratz LL, Marchini LC, Moreti ACCC. [Effect of Starter method and origin of larvae of Africanized honeybees (Apis mellifera L., 1758) in overlapped nucs for royal jelly production, in Portuguese]. Efeito do método 'Starter' e da origem das larvas de abelhas africanizadas (Apis mellifera L., 1758) em minirecrias para produção de geleia real. Boletim de Industria Animal 2002;59:71–77.

63. Muli EM, Raina SK, Mueke JM. Royal jelly production in East Africa: performance potential of the honey bees, Apis mellifera scutellata and Apis mellifera monticola in Kenya. Journal of Apicultural Research 2005;44:137–140. DOI: 10.1080/00218839.2005.11101167

64. Li JK, Feng M, Desalegn B, Fang Y, Zheng AJ. Proteome comparison of hypopharyngeal gland development between Italian and royal jelly producing worker honeybees (Apis mellifera L.). Journal of Proteome Research 2010;9:6578–6594. DOI: 10.1021/pr100768t

65. Su SK, Luo JN, Huang JX, Chen SL. Relationship between morphology of hypopharyngeal gland physiology development and royal jelly production of honeybee (Apis mellifera ligustica). Apiculture of China 2006;57:5–7.

66. Jianke L, Shenglu C, Boxiong Z, Songrun S. Optimizing royal jelly production. Good queens are a must! American Bee Journal 2003;143:221–223.

67. Bienefeld K, Ehrhardt K, Reinhardt F. Genetic evaluation in the honey bee considering queen and worker effects – a BLUP – animal model approach. Apidologie 2007;38:77–85. DOI: 10.1051/apido:2006050

68. Jianke L, Aiping W. Comprehensive technology for maximizing royal jelly production. American Bee Journal 2005;145:661–664.

69. Harbo JR, Rinderer TE. Breeding and genetics of honeybees. In: ARS Agriculture Handbook No. 335. Beekeeping in the United States. Washington, DC: USDA; 1980. p. 49–57.

70. Baitala TV, Nakasugui LF, Cantagalli LB, Toledo VAA, Ruvolo-Takasusuki MCC. Major royal jelly protein-3 polymorphism in Africanized honeybee colonies producing royal jelly. Magistra 2013;25:30–36.

71. Koeniger G, Koeniger N, Ellis J, Connor L. Mating biology of honey bees (Apis mellifera). Kalamazoo, Michigan: Wicwas Press, 2014. 155 p.

72. Woodward DR. Queen bee – biology, rearing and breeding. 2nd Ed. Balclutha, New Zealand: Woodward; 2009. 137 p.

03 ENHANCEMENT AND CULLING OF HONEYBEES FACILITATED BY MOLECULAR MARKERS

1. INTRODUCTION: AN OVERVIEW

The goal of genetic improvement in any organism is to enhance the frequencies of genes that are economically significant and need to be selected within the population. Regarding bees, this entails augmenting the frequency of colonies that generate offspring exceeding the average of the previous generation, from which the selection was conducted.

The production of royal jelly and honey is a collaborative effort of the workers in a colony. This makes the entire colony a unit of selection, where the evaluation of superior queens is done by the offspring of the workers involved in production. Studies on royal jelly production have revealed significant heterogeneity in its production by Africanized honeybees [3,4]. The results indicate the necessity of queen selection [5]. The process of selecting bees with specific genotypes requires utilizing advanced techniques such as enhanced queens' replacement methods, instrumental insemination, and molecular marker-assisted selection. There is a scarcity of literature data that establish a connection between molecular markers and the formation of royal jelly. The discovery and analysis of multiple loci containing the major royal jelly proteins (MRJPs) has enabled the utilization of one specific locus, Mrjp3, as a molecular marker for the purpose of selecting Africanized Apis mellifera queens. The process of choosing queens and genotyping the most productive ones commenced in 2006 at the apiary of the State University of Maringá, Brazil. The initial investigation linking the MRJP3 marker to the synthesis of royal jelly was conducted by [6]. The significant variation observed in the major royal jelly proteins (MRJPs), specifically MRJP3 which contains microsatellite regions, suggests that these proteins, particularly the microsatellite regions found in the Mrjp3 gene, have a high potential for being used as markers in studies aimed at enhancing royal jelly production. Following that, additional studies were carried out utilizing traditional performance indicators such as the MRJP3 marker. The current findings have demonstrated the significance of this marker in genotyping for the production of arrays of royal jelly. This chapter highlights the significance of royal jelly for both honey bees and human health. It emphasizes the need of enhancing the process through the use of molecular markers and presents the outcomes achieved by selecting queens that produce royal jelly and are genotyped for MRJP3.

2. ROYAL JELLY

Royal jelly is produced by the mandibular and hypopharyngeal glands situated in the head of honeybees [7]. The secretion from the hypopharyngeal gland is transparent and watery, including a high amount of protein. On the other hand, the secretion from the mandibular gland is white and has a milky consistency [7,8]. Royal jelly is a thick substance that is white-yellowish or grayish white in color. It has a somewhat opalescent appearance and a distinct pungent odor, although it is not unpleasant or rancid. (Figure 1) [9–10]. The development rate and activity of worker nurses in these glands peak between days 10 and 14 [11–15]. The development of the glands in a colony can be impacted by both internal factors, like as the presence of progeny and population density, and external events, such as foraging. These factors enable bees to adjust rapidly to the needs of the colony [13,16–18].

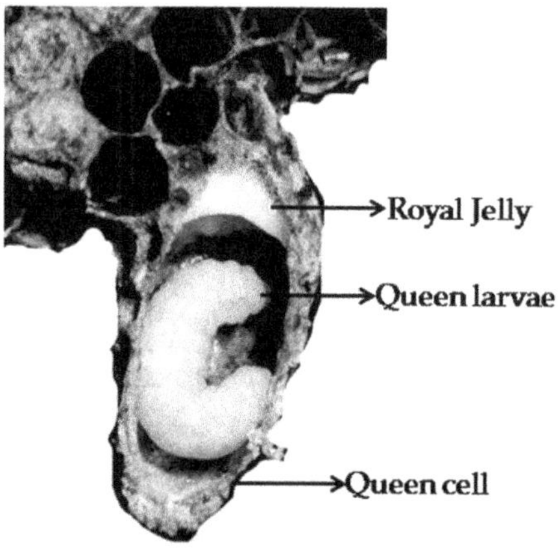

Figure 1: Queen cell with queen larvae of A. mellifera and royal jelly.

Royal jelly is a glandular secretion noted for its intricate composition, which includes minerals, proteins, amino acids, steroids, phenols, carbohydrates, vitamins, lipids, acetylcholine, and other unidentified chemicals [9,19]. It also plays a significant role in reproduction and development. During the first three days of growth, royal jelly serves as the primary source of nutrition for the larvae. After this period, it becomes the sole nourishment for the queen bee for the rest of her life. This exclusive diet ensures her fertility and promotes a longer lifespan. Starting on the third day, worker larvae are nourished with a combination of honey, pollen, and water, which is referred to as brood food. Drones, on the other hand, are provided with brood food and royal jelly [20].

The typical lifespan of A. mellifera queens is 1 to 2 years [21]. They reach sexual maturity 6 days after emerging and mate with approximately 17 drones. They store all the necessary sperm to fertilize eggs for their whole lifespan [22]. During the summer, there is a low number of drones being raised. However, there is a minor increase in drone rearing during swarming. Queens have the ability to produce between 1500 and 2000 eggs per day during their lifetimes [23,24]. The actual number of eggs laid depends on the requirements of the hive and many environmental circumstances. Meanwhile, a significant number of workers, who are sterile females, are responsible for the upkeep of the hive.

Studies have been performed to explore the potential effects of royal jelly, which is known for its ability to enhance reproduction and longevity in queens, on humans. The ingestion of royal jelly has been associated with several advantageous benefits, including the reduction of physical and mental exhaustion, the restoration of normal appetite, the stimulation of brain activity, enhanced vision, higher immunity against viral infections, and rejuvenation of the skin [10].

Due to its high content of proteins, free amino acids, lipids, vitamins, carbohydrates, and bioactive chemicals such 10-hydroxy-trans-2-decenoic acid and antibacterial protein 350 KDa proteins, royal jelly is used as a component in a variety of nutritious dishes [25]. A review conducted by [25] indicates that numerous research have documented the advantageous physiological and pharmacological effects of royal jelly in mammals. These effects include vasodilatory and hypotensive activities, antihypercholesterolemic activity, and anticancer activity.

3. THE MOLECULAR MARKER IS SIGNIFICANT. PROTEIN DERIVED FROM ROYAL JELLY

Royal jelly consists of soluble proteins in water and water-insoluble proteins, making up 12 to 15% of its crude protein content. The soluble proteins of royal jelly produced by the hypopharyngeal and mandibular glands consist primarily of large proteins with molecular weights ranging from 47 to 80 kDa, along with a modest quantity of smaller proteins such as antibiotics and peptides. These proteins are the primary constituents of the principal group of royal jelly proteins. The Major Royal Jelly Proteins (MRJPs) account for approximately 82 to 90% of the overall protein content in larval jelly [19]. Certain sections of MRJPs may prioritize amino acids that are nitrogen-rich, resulting in the accumulation of high quantities of nitrogen in MRJPs. Nitrogen availability is crucial for the rapid growth of young larvae and the development of the queen [29]. These observations provide evidence that MRJPs play a significant role in the nutritional needs of bees [19].

Major royal jelly protein-3 can be observed using denaturing SDS-PAGE electrophoresis in the head extracts of working nurses aged 10-14 days or in royal jelly, as shown in Figure 2. The degree of polymorphism was quantified by [30].

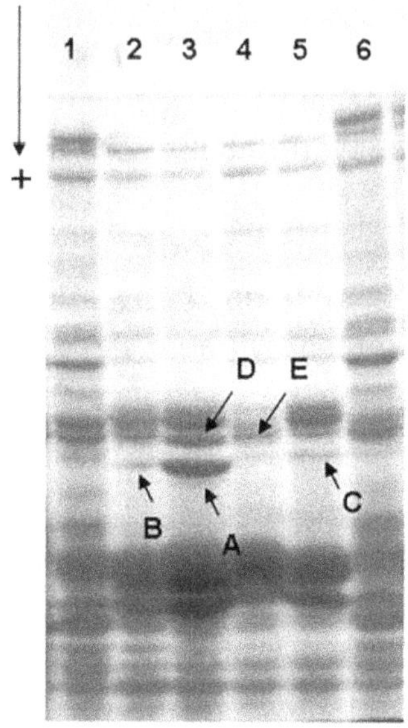

Figure 2: Denaturating SDS-PAGE electrophoresis showing MRJP3 polymorphism in extracts of head of A. mellifera nurse. A, B, C, D, E = alleles. Source: Baitala et al. (2013).

MRJPs genes encoding a group of proteins that have a common evolutionary origin with Yellow proteins of *Drosophila melanogaster* [31]. Genome of *Drosophila* encodes at least seven family members of Yellow proteins [32], whose loci are involved in the larval pigmentation [33], unlike the MRJPs that have nutritional function of larvae.

The genes encoding key proteins of royal jelly began to be identified in studies [34] and [35]. After these pioneering studies, several studies have been published in order to identify and characterize new genes encoding the MRJPs proteins [19,31,36,37]. The availability of the complete genome of *Apis mellifera* [38] made it possible to identify new genes encoding proteins of the family MRJP [29].

Since the first study were identified nine proteins in MRJPs *A. mellifera* (MRJP1, MRJP2, MRJP3, MRJP4, MRJP5, MRJP6, MRJP7, MRJP8, and MRJP9) besides an incomplete polypeptide, MRJPψ, encoded by a pseudogene. The genes encoding these proteins are located on chromosome 11 [29]. Classification of *A. mellifera* MRJPs has been performed based on the N-terminal sequences of purified protein and cDNA sequences available in the cDNA library.

The analysis conducted using PCR and DNA sequencing revealed that the alleles of the gene responsible for generating the MRJP3 protein exhibit variations in length due to the presence of a varied number of repeating basic units in a specific region of the Mrjp3 gene [31,36]. The authors ascribed the polymorphism of these proteins to the existence of a region containing a variable number of repeating sequences in tandem, also known as a microsatellite. These markers consist of a variable number of identical sequences, ranging from 15 to 100 base pairs, that are arranged in a consecutive and repeating manner, with a maximum repetition of 50 times. The molecular variations in four variants of MRJP3 have indicated that the diversity in these proteins is associated with the variety in size, which is genetically dictated by bees within the same colony [19].

The Mrjp3 locus is polymorphic and has been characterized by DNA sequencing, which has discovered five alleles. Additionally, PCR analysis has revealed at least 10 alleles of varying sizes [36]. This study also discovered a Mendelian pattern of inheritance and a significant amount of variation in the genomic locus of MRJP3.

While A. mellifera and other Apis bees with MRJPs have been studied, there is less information in the literature regarding the use of MRJPs as molecular markers for selecting and improving royal jelly production.

4. *APIS MELLIFERA* QUEENS' SELECTION USING MRJP3 MARKER

The objective of genetic improvement is to enhance the prevalence of favorable genes at economically significant loci within a population [44]. The objective of genetic bee breeding is to enhance the occurrence of colonies that surpass the average generation from which the selection was conducted. The process of selecting honeybees with superior genetic traits entails the implementation of advanced procedures such as enhanced queen replacement methods, artificial insemination, and molecular marker-assisted selection [45].

The process of selecting queens is conducted by genetic evaluation, which relies on estimating the (co)variance components and genetic parameters to identify bees with superior genetic traits. The heritability estimate for royal jelly production, as determined by Bayesian inference, was found to be 0.27% for acceptance, 0.10 for production per colony, and 0.55 per dome [5]. The authors' results demonstrated that queen selection can enhance colony production of royal jelly, larval acceptance, and dome production of royal jelly. Additionally, extrinsic influences have the potential to alter gene expression in individuals.

However, there is a scarcity of data in the existing literature about the correlation between molecular markers and the synthesis of royal jelly. A pioneering scientific study conducted by [46] aimed to identify DNA markers associated with the formation of royal jelly. The scientists examined a total of 96 alleles generated from 10 microsatellite loci. Based on the observed frequency of certain alleles, they were able to discover seven alleles that potentially serve as markers for bees that create significant amounts of royal jelly. Utilizing molecular markers, namely microsatellites, can aid in identifying genetic variations that may be valuable in distinguishing bee colonies with a high production of royal jelly.

The significant variability seen in MRJPs proteins, particularly in MRJP3 which contains a microsatellite region, indicates the promising possibility of using MRJP family proteins as markers for selecting queen bees that can enhance royal jelly output. The utilization of MRJPs as molecular markers in population genetics research and as selection markers for enhancing royal jelly output remains limited. Several researchers have demonstrated the efficacy of this molecular marker for selecting royal jelly-producing queens.

The Africanized honeybees that were specifically chosen for royal jelly production had a significant amount of genetic variation at the Mrjp3 locus (Figure 3), indicating the potential of this marker for selective breeding [6]. This research involved doing evaluations of several linear regressions using EPD (expected progeny differences) values to predict royal jelly production. The variance studies revealed that the repetitive region of Mrjp3 had a significant impact on the genetic worth of the queen's children in terms of royal jelly production. The coefficient of determination (R2) for the important alleles of the repetitive region of Mrjp3 indicates that 36.85% of the variation in EPD can be explained by the variation in the C, D, and E alleles. The authors determined that the three alleles had a significant genetic impact on the variation of royal jelly output.

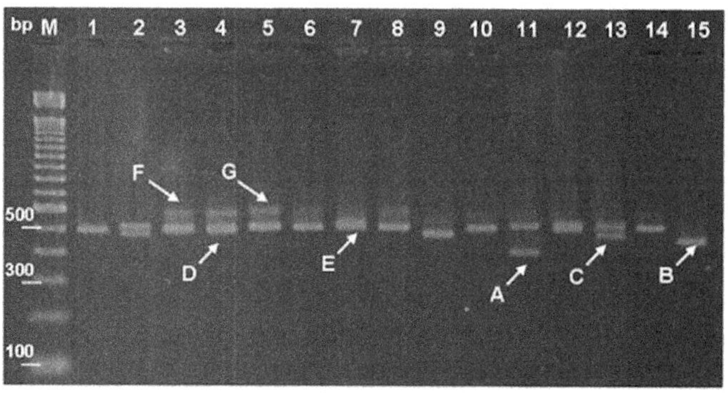

Figure 3: Molecular marker MRJP3. Number = A. mellifera DNA. A, B, C, D, E, F, G = Mrjp3 alleles. M = molecular weight marker. Source: Baitala et al (2010).

Continuing the process of selection and the Program of Improvement of *Apis mellifera* Royal Jelly Producing (PIAMRJP), State University of Maringá, Brazil, alleles of the locus *Mrjp3* descendants queens, those selected by [6], were evaluated in 2011 [47]. Results showed that the royal jelly-producing queens had a high degree of genetic diversity and excess homozygous alleles. The highest frequencies were estimated for *Mrjp3 D* and *E* alleles 0.3357 and 0.3107, respectively, showing that the selection process of queens royal jelly producing these alleles are being maintained and only the *C* allele had a low frequency of 0.0321.

Results obtained by [47] confirm those obtained by [6], the locus *Mrjp3* and their alleles *C*, *D* and *E* influence the genetic value for producing royal jelly; however, the real role of MRJP3 these bees has not yet been identified. The sequencing of *Mrjp3* of *A. mellifera* Africanized alleles in PIAMRJP was performed [48]. Homology and identity of these sequences were compared with the sequences deposited in the database for *A. mellifera* (Figure 4). Alleles *Mrjp3* detected showed high identities with alleles deposited in BLAST system. Alleles *Mrjp3 C, D* and *E* are being maintained in the genome of the selected matrices queens. High similarity among the *Mrjp*3 alleles analyzed and those described in other studies show that the *Mrjp3* locus is conserved among species and subspecies of *Apis*. Similar results were obtained by [40]. These authors found that there are high similarity sequences and intron-exon have the same structure between four species *A. mellifera, A. cerana, A. dorsata* and *A. florea.* The procedure of selecting royal jelly-producing queens can lead to the selection of reproduction bees with specific genetic features, which can then be passed down from one generation to the next [49]. Nevertheless, it is crucial to retain a certain level of genetic diversity, since it leads to a greater capacity for selection enhancement [50]. Furthermore, with the ongoing process of identifying the genetic makeup of queens who produce royal jelly, we conducted a study to determine the origin of the mitochondrial DNA (mtDNA) in Africanized bees A. mellifera that are part of the breeding program. Specifically, we aimed to ascertain if their mtDNA can be traced back to African or European origins. [51] conducted this research utilizing matrices that generate royal jelly.

The analysis of mitochondrial DNA in European and African honey bees was conducted using the molecular marker PCR-RFLP, employing particular primers and restriction enzymes. In 2013, analyses were conducted on the daughters of queens that produce royal jelly. These queens were part of the PIAMRJP project, which began in 2006 and had been ongoing for seven years at the time of the analyses. Following a thorough process of selecting and analyzing genetic factors, alleles C, D, and E are currently being preserved in queens, highlighting their significant contribution to the generation of royal jelly. Queens chosen for the manufacture of royal jelly exhibited a higher prevalence of African mitochondrial DNA (mtDNA), indicating that the genes inherited from the mother are of African origin. The utilization of microsatellite markers and mtDNA can be employed in bee enhancement initiatives to ascertain the genetic lineage of queens and validate the effectiveness of the Program of Improvement of Apis mellifera Royal Jelly Producing [51].

Utilizing molecular markers in selection programs and honey bee improvement for royal jelly production is effective as it enables the preservation of desired genotypes, ensuring optimal hive productivity. The microsatellite marker MRJP3 has proven to be an effective tool for verifying the genotypes of breeding queens, making it easier to identify and manage beehives in the apiary of the Royal Jelly Producing Program for Apis mellifera.

```
A.m.protein                 ------------------------------------------------CAATCAGAATGCT 13
A.m.protein3                ------------------------------------------------CAATCAGAATGCT 13
A.m.protein3-like           ----------------------------CTGGCAATCAGAATGCTGGCAATCAGAATGCT 32
A.m.carnicaprotein3         ---------------------GGAAGATATCACAATCAGAATGCTGGCAATCAGAATGCT 39
MRJP3-C                     ATTATCATTTTGCCTGTTTACCATTCCTCTTGTTATCATTCTGTCTGTTACCATTTTGCT 60
                                                                             * **   ****

A.m.protein                 GGCAATCAGAATGCTGACAATCA----GAATGCTGACAATCAGAATGCTAACAATCAGAA 69
A.m.protein3                GGCAATCAGAATGCTGACAATCA----GAATGCTGACAATCAGAATGCTAACAATCAGAA 69
A.m.protein3-like           GGCAATCAGAATGCTGACAATCA----GAATGTTGACAATCAGAATGCTAACAATCAGAA 88
A.m.carnicaprotein3         GGCAATCAGAATGCTGACAATCA----GAATGCTGACAATCAGAATGCTAACAATCAGAA 95
MRJP3-C                     TGTTATCATTTTGTCTATTACCATTTTGCTTGTTATCATTCTGTTTGTTACCATTTTGTC 120
                             *  ****   **   *  * **    *  ** *  ** ** *  ** ** ** *  *

A.m.protein                 TGCTGATAATCAGAATGCTAACAAACAAAATGGTAATAGACAAAATGATAACAGACAGAA 129
A.m.protein3                TGCTGATAATCAGAATGCTAACAAACAAAATGGTAATAGACAAAATGATAACAGACAGAA 129
A.m.protein3-like           TGCTGATAATCAGAATGCTAACAAACAAAATGGTAATAGACAAAATGGTAACAGACAGAA 148
A.m.carnicaprotein3         TGCTGATAATCAGAATGCTAACAAACAAAATGGTAATAGACAAAATGATAACAGACAGAA 155
MRJP3-C                     TGTTACCATTTTGCTTGTTATCATTCTGTCTGTTACCATTTTGCTTGTTATCATTCTGTC 180
                            ** *   * *  *  ** ** **  *    ** ** **  *       ** ** **  * *

A.m.protein                 TGATAACAAGCAAAATGGTAACAGACAGAATGATAACAAGCAAAATGGTAACAGACAGAA 189
A.m.protein3                TGATAACAAGCAAAATGGTAACAGACAGAATGATAACAAGCAAAATGGTAACAGACAGAA 189
A.m.protein3-like           TGATAACAAGCAAAATGGTAACAGACAGAATGATAACAAGCAAAATGGTAACAGACAGAA 208
A.m.carnicaprotein3         TGATAACAAGCAAAATGGTAACAGACAGAATGATAACAAGCAAAATGGTAACAGACAGAA 215
MRJP3-C                     TGTTACCATTTTGCTTGTTATCATTCTGTCTGTTACCATTTTGCTTGTTATCATTCTGTC 240
                            ** ** **       ** ** **  * *  ** ** **       ** ** **  * *

A.m.protein                 TGATAACA--------AGCAAAAT-------GGTAACAGACAAAATGGTAACAA------ 228
A.m.protein3                TGATAACA--------AGCAAAAT-------GGTAACAGACAAAATGGTAACAA------ 228
A.m.protein3-like           TGATAACA--------AGCAAAAT-------GGTAACAGACAGAATGATAACAAGCAAAA 253
A.m.carnicaprotein3         TGATAACA--------AGCAAAAT-------GGTAACAGACAGAATGATAACAAGCAAAA 260
MRJP3-C                     TGTTATCATTTTGTCTATTACCATTTTGTTTGTTAGCATTCTGATTATCAGCATTCTGAT 300
                            ** ** **        *  *  **       * ** **  *  * *   * **

A.m.protein                 ---------ACAGAATGATAACAAGCAAAATGGTAACAGACAGAATGATAACAAGAGGAA 279
A.m.protein3                ---------ACAGAATGATAACAAGCAAAATGGTAACAGACAGAATGATAACAAGAGGAA 279
A.m.protein3-like           TAGTAACAGACAGAATGATAACAAGCAAAATGGTAATAGACAAAATGGTAACAAACAGAA 313
A.m.carnicaprotein3         TGGTAACAGACAGAATGATAACAAGCAAAATGGTAACAGACAAAATGGTAACAAACAGAA 320
MRJP3-C                     TGTTAGCATTCTGATTGTCAGCATTCTGATTGTCAGCATTCTGATTGCCAGCATTCTGAT 360
                                      * ** **  * **  *  * **  *  *  *  * **  * **    **

A.m.protein                 TGGTAACAGGCAAAATGATAAT-------------CAA---------------------- 304
A.m.protein3                TGGTAACAGGCAAAATGATAAT-------------CAA---------------------- 304
A.m.protein3-like           TGATAACAAGCAAAATGGTAATA----------GACAAAATGATAACAAGCAAAATGGTA 363
A.m.carnicaprotein3         TGATAACAAGCAAAATGATAATA----------GACAAAATGATAACAAGCAAAATGGTA 370
MRJP3-C                     TGCCAGCATTCTGATTGCCAGTCTCTTCCGGTTGCCGAGCTGTT--CCTGCATTGTGATC 418
                            **  * **  *  * **  * *             * *

A.m.protein                 ----------------------------------------------
A.m.protein3                ----------------------------------------------
A.m.protein3-like           ACAGACAGAATGATAA------------------------------ 379
A.m.carnicaprotein3         ACAGACAGAATGATAACAAGAGGAATGGTAACAGGCAAAATGATAAT 417
MRJP3-C                     ACTTCCCGGAAGGGA------------------------------- 433
```

Figure 4: Alignment of sequences similar to the Mrjp3 C allele performed using ClustalW2 (EMBL-EBI); sequences include A. mellifera major royal jelly protein mRNA, complete cds (GU434675.1); A. mellifera major royal jelly protein 3 (Mrjp3), mRNA (NM_001011601.1); A. mellifera carnica major royal jelly protein 3 (Mrjp3) gene, complete cds (AY663104.1); and PREDICTED: A. mellifera major royal jelly protein 3-like (LOC727045), partial mRNA (XM_001122757.2). "*" = nucleotides identical in all of the aligned sequences. Source: Casagrande-Pozza (2011).

REFERENCES

1. Collins A.M.; Rinderer T.E.; Harbo J.R.; Brown M.A. Heritabilities and correlations for several characters in the honey bee. The Journal of Heredity, 1984;75:135–140.
2. Rinderer T.E. Measuring the heritability of characters of honeybees. Journal of Apicultural Research, 1977;16:95–98.
3. Mouro G.F.; Toledo V.A.A. Evaluation of Apis mellifera Carniolan and Africanized honey bees in royal jelly production. Brazilian Archives of. Biology and Technology. 2004;47:469–476.
4. Garcia R.H.C.; Nogueira-Couto R.H. Produção de geléia real por abelhas Apis mellifera italianas, africanizadas e descendentes de seus cruzamentos. Acta Scientiarum Animal Sciences, 2005;27(1):17–22.
5. Faquinello P.; Toledo V.A.A.; Martins E.N.; Oliveira C.A.L.; Sereia M.J.; Costa-MaiaF. M.; Ruvolo-Takasusuki M.C.C. Parameters for Royal Jelly Production in Africanized Honeybees. Sociobiology, 2011;57(3):495–509.
6. Baitala T.V., Faquinello P.; Toledo V.A.A.; Mangolin C.A.; Martins E.N.; Ruvolo-Takasusuki M.C.C. Potential use of major royal jelly proteins (MRJPs) as molecular markers for royal jelly production in Africanized honeybee colonies, Apidologie, 2010;41:160–168.
7. Haydak M.H. Honey bee nutrition. Annual Review of Entomology, 1970;15:143–156.
8. Beetsma J. The process of queen-worker differentiation in the honeybee. Bee World, 1979;60(1):24–39.
9. Garcia-Amoedo L.H.; Almeida-Muradian L.B. Physicochemical composition of pure and adulterated royal jelly. Química Nova, 2007;30(2):257–259.
10. Koshio S.; Almeida-Muradian L.B. de. Aplicação da CLAE para determinação do ácido 10-hidróxi-2-decenóico (10-HDA) em geléia real pura e adicionada a mel brasileiro. Química Nova, 2003;26(5):670–673.
11. Takenaka T.; Kaatz H.H. Protein synthesis by hypopharyngeal glands of worker honey bees. In: Eder J; Rembold H (eds.). Chemistry and Biology of Social Insects. München: Peperny, 1987. pp. 166–167.
12. Crailsheim K.; Stolberg E. Influence of diet, age and colony condition upon intestinal proteolytic activity and size of the hypopharyngeal glands in the honeybee (Apis mellifera L.). Journal of Insect Physiology, 1989;35:595–602.
13. Huang Z.Y.; Otis G.W. Factors determining hypopharyngeal gland activity of worker honey bees. Insectes Sociaux, 1989;36(4):264–276.

14. Knecht D.; Kaatz H. H. Patterns of larval food production by hypopharyngeal glands in adult worker honey bees. Apidologie, 1990;21:457–468.

15. Naiem E.S.; Hrassnigg N.; Crailsheim K. Nurse bees support the physiological development of young bees (Apis mellifera L.). Journal of Comparative Physiology, 1999;169:271–279.

16. Simpson J.; Riedel I.B.M.; Wilding N. Invertase in the glands hypopharyngeal of the honeybee, Journal of Apicultural Research, 1968;7:29–36.

17. Brouwers E.V.M. Measurement of hypopharyngeal gland activity in the honeybee, Journal of Apicultural Research, 1982;21:193–198.

18. Silva de Moraes R.L.M.; Cruz-Landim C. Influência da densidade populacional no comportamento dos núcleos das glândulas hipofaringeanas de Apismellifera (Hymenoptera, Apidae). Naturalia, 1984;9:27–33.

19. Schmitzová J.; Klaudiny J.; Albert S.; Schröder W.; Schreckengost W.; Hanes J.; Júdová J.; Simúth J. A family of major royal jelly proteins of the honeybee Apis mellifera L. Cellular and Molecular Life Sciences, 1998;54:1020–1030.

20. Winston M. The biology of the honey bee. Cambridge: Harvard University Press, 1991. 294 p.

21. PageRE; PengCYS Aging and development in social insects with emphasis on the honey bee, Apis mellifera L. Experimental Gerontology, 2001;36:695–711.

22. Woyke J. Natural and artificial insemination of queen honey bees. Pszczelnicze Zeszyty Nauwoke, 1960;4:183–273.

23. Merrill J.H. Observations on brood-rearing. American Bee Journal, 1924;64:337–338.

24. Nolan WJ. The brood-rearing cycle of the honey bee. Bulletin of U.S. Department of Agriculture, 1925;1349:1–56.

25. Oršolić N. Royal jelly: component efficiency, analysis, and standardization. Arh Hig Rada Toksikol. 2013;64(3):445–461. doi: 10.2478/10004-1254-64-2013-2332.

26. Hanes J.; Smúth J. Identification and partial characterization of the major jelly protein of the honeybee (Apis mellifera L.). Journal of Apicultural Research, 1992;31:22–26.

27. Fujiwara S.; Imai J.; Fugiwara J.; Yaeshima T.; Kobayashi K.A. potent antibacterial protein in a royal jelly. Journal of Biological Chemistry, 1990;265:11333–13337.

28. Biliková K.; Wu G.; Simúth J. Isolation of peptide fraction from honeybee royal jelly as anti faul brood factor. Apidologie, 2001;32:175–283.

29. Drapeau M.D.; Albert S.; Kucharski R.; Prusko C.; Maleszka R. Evolution of the Yellow/Major royal jelly protein family and the emergence of social behavior in honey bees. Genome Research, 2006;16:1385–1394.

30. Baitala T.V.; Nakasugui L.F.; Cantagalli L.B.; Toledo V.A.A.; Ruvolo-Takasusuki M.C.C. Major royal jelly protein-3 polymorphism in Africanized honeybee colonies producing royal jelly. Magistra, 2013;25(1):30–36.

31. Albert S.; Bhattacharya D.; Kaludiny J.; Schmitzová J.; Simúth J. The family of major royal jelly proteins and its evolution. Journal of Molecular Evolution, 1999a;49:290–297.

32. Maleszka R.; Kucharski R. Analysis of Drosophila yellow-B cDNA reveals a new family of proteins related to the royal jelly proteins in the Honeybee and to an orphan protein in an unusual bacterium Deinococcus radiodurans. Biochemical Biophysics and Research Communications, 2000;270(3):773–776.

33. Walter M.F.; Black B.C.; Afshar G.; Kermabon A.Y.; Wrigth T.R.F.; Biessmann H. Temporal and spatial expression of the yellow gene in correlation with cuticle formation and DOPA decarboxylase activity in Drosophila development. Developmental Biology, 1991;147:32–45.

34. Klaudiny J.; Hanes J.; Kulifajova J.; Albert S.; Simúth J. Molecular cloning of two cDNAs from the head of the nurse honey bee (Apis mellifera L.) coding for related proteins of royal jelly. Journal of Apicultural Research, 1994;33:105–111.

35. Albert S.; kaudiny J.; simúth J. Newly discovered features of the updated sequence of royal jelly protein RJP57-1; longer repetitive region on C-terminus and homology to Drosophila melanogaster yellow protein. Journal of Apicultural Research, 1996;35:63–68.

36. Albert S.; Klaudiny J.; Simúth J. Molecular characterization of MRJP3, highly polymorphic protein of honeybee (Apis mellifera) royal jelly. Insect Biochemistry and Molecular Biology, 1999b;29:427–434.

37. Albert S.; Klaudiny J. The MRJP/YELLOW protein family of Apis mellifera: Identification of new members in the EST library. Journal of Insect Physiology, 2004;50:51–59.

38. The Honey Bee Genome Sequencing Consortium. Insights into social insects from the genome of the honeybee Apis mellifera. Nature, 2006;443(26):931–49.

39. Albert S.; Schmitz J. Characterization of major royal jelly protein-like DNA sequences in Apis dorsata. Journal of Apicultural Research, 2002;41(3–4):75–85.

40. Srisuparbh D.; Klinbunga S.; Wongsiri S.; Sittipraneed S. Isolation and Characterization of Major Royal Jelly cDNAs and Proteins of the Honey Bee (Apiscerana). Journal of Biochemistry and Molecular Biology, 2003;36(6):572–579.

41. Albertová V.; Su S.; Brockmann A.; Gadau, J; Albert S. Organization and Potential Function of the mrjp3 Locus in Four Honeybee Species. Journal of Agricultural and Food Chemistry, 2005;53:8075–8081.

42. Imjongjirak C.; Klinbunga S.; Sittipraneed S. Cloning, expression and genomic organization of genes encoding major royal jelly protein 1 and 2 of the honey bee (Apis cerana). Journal of Biochemistry and Molecular Biology, 2005;38(1):49–57.

43. Su S.; Albert S.; Chen S.; Zhong B. Molecular cloning and analysis of four cDNAs from the heads of Apis ceranacerana nurse honeybees coding for major royal jelly proteins. Apidologie, 2005;36:389–401.

44. Bergmann J.A.G. Avaliação genética. In: PEREIRA J.C.C. (eds.). Melhoramento genético aplicado à produção animal. 3. ed. Belo Horizonte: FEPMVZ, 2001. pp. 502–515.

45. Harbo J.R.; Rinderer T.E. Breeding and genetics of honeybees. Beekeeping in the United States. 2005. Disponível em: http://maarec.cas.psu.edu/bkCD/HBBiology/breeding_genetics.htm. Acess in 11/19/2007.

46. Chen S; LI J.; Zhong B.; Su S. Microsatellite analysis of royal jelly producing traits of Italian honeybee (Apis mellifera Liguatica). Acta Genetica Sinica, 2005;32(10):1037–1044.

47. Parpinelli R.S.; Ruvolo-Takasusuki M.C.C.; Toledo V.A.A. MRJP microsatellite markers in Africanized Apis mellifera colonies selected on the basis of royal jelly production. Genetics and Molecular Research, 2014;13(3):6724–6733. DOI: http://dx.doi.org/10.4238/2014.August.28.16

48. Casagrande-Pozza A.P.B. Sequenciamento dos alelos das proteínas principais da geléia real de abelhas Apis mellifera africanizadas. UEM: Universidade Estadual de Maringá, Maringá, 2011.52 p. Dissertação (Mestrado em Zootecnia).

49. Falconer D.S. (1987) Introdução à genética quantitativa, Universidade Federal de Visçosa, Minas Gerais.

50. Page R.E., Kerr W.E. Honey bee genetics and breeding, in: Spivak M., Fletcher D.J.C., Breed M.D. (Ed.), The "African" honey bee, Colorado: Westview Press, 1991; pp.157-186.

51. Oliveira A.P.N.Z. Seleção de abelhas Apis mellifera L. Africanizadas produtoras de geleia real por marcadores moleculares. UEM: Universidade Estadual de Maringá, Maringá, 2012. 41p. Dissertação (Mestrado em Genética e Melhoramento).

04 EFFECTS OF PESTICIDES ON HONEY BEES

1. INTRODUCTION: AN OVERVIEW

Beekeepers have long been aware of the environmental factors that contribute to the success of their honey bee colonies, such as a wide variety of blooms from various sources including trees, shrubs, so-called weeds, and even crop plants. An abundant and varied floral environment has always been the key factor for achieving a high and successful honey output. The only challenges they encountered were sporadic microbial infections, illnesses, and parasites that had the potential to exterminate the bees and their colonies [1], as well as the unexpected fluctuations in weather that could negatively impact flower yield during particularly unfavorable years.

Recently, beekeepers have faced a new challenge to their industry: the use of agrochemical pesticides. These pesticides are applied across extensive areas of crops, fruit groves, woods, and other habitats to manage insect pests, weeds, vermin, and plant diseases. Undoubtedly, chemical insecticides pose a significant risk to bees due to their susceptibility as insects to any toxic substance intended to eliminate insect pests. As a result, rigorous toxicity testing[2, 3] has been and continues to be necessary prior to the approval of these pesticides for use in crop protection. This need is particularly applicable in industrialized nations. Notwithstanding these rules, the population of controlled honey bee (Apis mellifera L.) colonies in the United States decreased from 6 million in 1947, when DDT was used in agriculture, to less than 3 million in 2010 [4]. In Europe, there has been a decrease in the number of apiaries by 14% in Scandinavia and 25% in central Europe from 1985 to 2005. However, there has been a 13% growth in apiaries in Mediterranean countries to compensate for the reduced production in the northern regions.

However, what about the presence of other types of pesticides, such as herbicides and fungicides? Could they also impact honey bee productivity? Some have suggested that if the pesticides do not target insects, they are likely to be harmless for bees. Recent research undertaken in countries with a significant historical use of pesticides indicates otherwise. It is now recognized that the widespread and long-term use of herbicides results in a decrease in the variety of flowering plants [6, 7], which ultimately impacts the colonies of bees [8] and their ability to produce. Furthermore, research has shown that the combination of certain fungicides and insecticides is more lethal to bees than either chemical used alone [9]. Recently, the widespread use of acaricides in bee farms to combat parasites like Varroa destructor has become an additional concern for beekeepers, as these chemicals are also harmful, albeit to a lesser extent, to honey bees. It is not unexpected that some authors have linked colony collapse disorder (CCD) not just to parasites and illnesses, but also to the use of pesticides [10].

Given the current circumstances, it is imperative to use a novel management technique in order to achieve success in the field of beekeeping. The production of honey and wax is no longer reliant on the presence of flowers in the nearby environment. Instead, it seems to be closely connected to the nutritional value of the food that the bees gather. Evidence has emerged indicating that flowers tainted with pesticides have a detrimental impact on the health of honey bee colonies, leading to a fall in their output [11]. To effectively handle this issue, it is crucial to gain a comprehensive understanding of the mechanisms via which bees come into contact with pesticides and the resulting implications for the health of individual bees, the colony, and their overall production.

2. EXPOSURE OF BEES TO AGROCHEMICALS

Insecticides are typically sprayed above the crop canopy, while herbicides and fungicides are frequently applied directly to the soil prior to crop planting. In each of these instances, it is possible for droplets and dust emitted from the treatments to come into direct contact with bees that are flying over the treated fields or in close proximity. This is because the wind has the ability to transport these minuscule droplets and dust particles over distances of hundreds of meters from the crop [12]. A solitary droplet of insecticide can be lethal to a bee due to the concentrated dosages of chemicals present in the spray solutions. This is the primary reason for the bee-related occurrences documented in the literature [13, 14]. Bees are not directly exposed to granular pesticides, such as herbicides, that are mixed into the soil.

The so-called systemic insecticides are usually applied as seed coatings. The treated seeds are introduced into the soil using pneumatic drilling planters, and the friction of the seeds in the machinery produces dust particles that are heavily loaded with the insecticides. These poisonous particles can also cause a great deal of mortality among bees, if they happen to be in the surroundings [15]. Systemic insecticides applied this way are taken up by the crop plants as they grow and their residues are present in all parts of the treated plant, including the flowers, pollen and nectar [16]. Not only the crop plants but also the weeds and bushes that grow in the vicinity are affected [17, 18] because they also take up small amounts of residues that spread through the soil through lateral water flow [19] or are contaminated through dust/spray drift. In addition, some plants can produce guttation drops in the early hours of the morning (e.g. maize, strawberries), and systemic insecticides appear in such drops in elevated concentrations [20] that are capable of killing the bees.

Most of the time, the exposure of bees to pesticides is through ingestion of residues found in the pollen and nectar of contaminated plants, whether from the crop plants or from the weeds around the fields [21]. It is important to realise that bees forage everywhere they can and search for the most suitable flowers that produce pollen and nectar in abundance. Thus, some crops are more attractive than others; for example, the yellow flowers of canola (rape seed oil), sunflowers and many weeds that grow in and around the crops are more attractive to bees than the flowers of potato plants. Pesticide residues in pollen and nectar are taken by the forager bees to their colonies and remain in the beebread and honey for quite some time [22, 23]. These residues are then fed to the larvae and the queen, which are affected in similar ways as the forager bees.

In addition to food, bees also drink water to keep their body temperature under control [24]. Pesticide residues in soil eventually move into the water and appear in the streams, creeks and ponds of agricultural areas and beyond, which are thus contaminated with a mixture of agrochemicals [25]. Some water contamination is also due to drift from spray applications, particularly from insecticides [26, 27]. Honey bees, bumblebees and wild bees like to drink from puddles, irrigation ditches, ponds and streams, and if these waters are contaminated with pesticide residues, the forager bees ingest them as well [28].

In addition to the pesticides utilized in agricultural production, honey bees are also subjected to acaricides employed for the management of Varroa and other parasites. Bees are exposed to high levels of residue found on the waxy cells of the comb [29]. This mostly affects the growth of the larvae and likely has an impact on adult honey bees and the queen.

Figure 1. Routes of exposure of bees to agricultural pesticides.

Due to the extensive range of agrochemicals utilized in agricultural cultivation, it is unsurprising that residues of 173 distinct substances have been detected in apiaries thus far [21]. It is important to understand that bees are not only threatened by one or two chemicals, but by mixtures of multiple agricultural substances, as they are exposed to pesticides through various pathways in the environment (Figure 1).

3. TOXICOLOGY OF PESTICIDES

Pesticides are chemical substances that are poisonous and have a specific mode of action. This means that they are specially designed to regulate a particular group of organisms by interfering with specific metabolic processes. Insecticides and acaricides eliminate insects and mites by interfering with their neural activity, moulting process, or other specific metabolic functions. Herbicides and algicides eradicate plants and algae by disrupting their photosynthetic capabilities or the synthesis of vital organic compounds. Fungicides eradicate fungi by inhibiting the formation of their cell membranes or other specific metabolic processes. Additionally, there are various types of substances such as rodenticides, which are designed to eliminate tiny mammals, as well as bird repellents, among others. The term biocide is specifically used to refer to potent substances that have the ability to kill a wide range of organisms, including bacteria and even larger animals.

The toxicity of each type of pesticide is not limited to the target group of organisms alone. Other species with comparable metabolism are also impacted, albeit usually to a lower extent. The toxicity of a pesticide to a particular species is determined by the amount of poisonous chemical that is fatal to 50% of individuals in that species (LD50), and this amount differs among different species. Doses below the LD50 threshold are classified as 'sublethal', but they can nevertheless result in mortality for a specific percentage of the species population, such as 20 or 30% of individuals. Typically, sublethal levels induce toxic effects that do not result in the death of organisms, but nonetheless disrupt their normal functioning and overall health. For instance, when bees are exposed to small amounts of neurotoxic pesticides, it can lead to stress [31], paralysis, or aberrant behaviors, without causing the death of the bees [32].

Insecticides are inherently highly harmful to bees, but herbicides are often harmless to them (Table 1). Beekeepers should exercise caution over any insecticide treatment near their hives, since spray drift has the potential to do significant harm to the bees. Pesticide applicators possess knowledge of this hazard and, in numerous nations, are obligated to notify beekeepers prior to administering insecticides to a crop [33]. Furthermore, although acaricides are less harmful to bees compared to the parasites they are intended to target, an excessive presence of acaricide residues in the combs might have negative effects on the bees' health [34].

Table 1. Toxicity of common pesticides to bees (LD50 at 48 hours) by contact or oral exposure and their persistence in soil (half-life)

Pesticide type	Chemical name	Contact* LD50(μg/bee)	Oral* LD50 (μg/bee)	Half-life† (days)
Acaricides	Acrinathrin	0.17	0.12	22
	Amitraz	50	-	1
	Coumaphos	20	4.6	-
	Fenpyroximate	11	-	49
	Tau-fluvalinate	8.7	45	4
	Tetradifon	1250	-	112
Fungicides	Azoxystrobin	200	25	78
	Boscalid	200	166	118
	Captan	215	91	4
	Carbendazim	50	-	22
	Chlorothalonil	135	63	44
	Myclobutanil	40	34	35
	Propiconazole	50	77	214
	Quintozene	71	-	210
	Tebuconazole	200	83	47
Herbicides	Metolachlor	-	110	90
	Norflurazon	1485	-	225
	Simazine	879	-	90
Insecticides	Beta-cyfluthrin	0.031	0.050	13
	Bifenthrin	0.015	0.20	87
	Carbofuran	0.16	-	14
	Chlorpyrifos	0.072	0.24	50
	Clothianidin	0.039	0.004	121
	Cypermethrin	0.034	0.064	69
	DDT	8.8	5.1	6200
	Diazinon	0.38	0.21	18
	Endosulfan	6.4	21	86
	Fenthion	0.22	-	22
	Fipronil	0.007	0.001	142
	Imidacloprid	0.061	0.013	174
	Malathion	0.47	9.2	1
	Mevinphos	0.094	-	1
	Pyrethrum	0.18	0.057	-

*Source: ECOTOX and AGRITOX databases, compiled in Sánchez-Bayo and Goka [21].

†Source: Footprint database (IUPAC). http://sitem.herts.ac.uk/aeru/iupac/

All species, including bees, possess detoxification processes that convert and remove the majority of harmful substances. Presently, most organic insecticides may be broken down either within the organisms themselves or in the surrounding environment. Organochlorine pesticides, such as DDT and lindane, are an exception since they are highly persistent and resistant to degradation. Due to their extensive application in previous decades, the residues of these substances nevertheless persist in the soils of numerous countries, albeit at low concentrations, despite being currently prohibited for use in agriculture. Because organochlorine residues have limited solubility in water, plants growing in polluted soils do not absorb them. As a result, these residues are not present in the pollen or nectar of the flowers.

The assessment of pesticide persistence is determined by its half-life (t1/2), which is the duration needed for half of the quantity of a chemical to degrade in a given medium, such as water, soil, air, or biological tissues. Pesticides with half-lives exceeding 90 days suggest the potential for accumulation, as over 5% of the initial amount applied will persist in the environment after one year [35]. Persistent insecticides residues detected in pollen or nectar (as shown in Table 1) will persist in the beebread throughout the whole honey production season.

Neonicotinoids (e.g. imidacloprid) and fipronil are examples of systemic insecticides that are more toxic and long-lasting compared to most organophosphorus (e.g. malathion), carbamates (e.g. carbofuran), and pyrethroids (e.g. cypermethrin) as shown in Table 1. Due to their significant solubility in water, these substances can be found in water bodies located in agricultural regions as well as in the rivers that receive runoff from these areas [36, 37]. When frequently used as seed dressings, these substances can persist in the soil for extended periods of time. They can be absorbed by the crops and weeds, ultimately ending up in the nectar and pollen of all plants in the treated area [16]. Bees are at risk due to the combination of their high toxicity, widespread availability, and unique method of action. Neonicotinoids, such as the ones mentioned, exhibit delayed toxicity when administered in small amounts. In addition to causing different sublethal effects, these substances ultimately result in the death of bees if they are exposed to the residues over an extended period of time. Both neonicotinoids and fipronil have the ability to weaken the immune system of honey bees, making them more susceptible to Nosema infections and viral diseases that are often spread by Varroa mites. Consequently, colonies that consume honey and pollen polluted with these neurotoxic insecticides may die due to the combined impact of chemicals and illnesses [45].

The presence of ergosterol-inhibiting fungicides, such as propiconazole and myclobutanil, might increase the toxicity of certain insecticides. These fungicides work as synergists. These chemicals effectively hinder the detoxification mechanism in bees [46, 47], resulting in a slower metabolism and elimination of pesticide and acaricide residues. Moreover, the harmfulness of insecticides and acaricides employed to manage Varroa mites is sometimes cumulative or mutually enhancing [9]. The collection of food by forager bees typically contains a combination of insecticides and fungicides, which are often contaminated. Additionally, acaricides are commonly used in managed apiaries. The cumulative toxicity and synergistic effects of these chemicals present a genuine danger to the well-being and survival of honey bee colonies as well as other wild bee species.

Animals often experience stress when they are exposed to pesticides, such as fungicides and some herbicides, at levels that do not immediately kill them. This is because the organisms attempt to swiftly metabolize and eliminate the poisonous compounds, which requires a significant amount of energy. In addition to stress, bees encounter numerous adverse consequences when exposed to sublethal quantities of insecticides. As an illustration, when honey bee larvae are consistently exposed to chlorpyrifos-contaminated pollen, they produce a minimal number of queens [48]. Wild bees, specifically Osmia bicornis, that were exposed to low levels of thiamethoxam and clothianidin experienced a 50% decrease in their ability to reproduce [49]. Honey bee queens, on the other hand, had an unusually high rate of supersedure, with 60% being replaced [50]. Bumble bee colonies, specifically Bombus terrestris, that were exposed to low levels of thiamethoxam were unable to function properly and produced 85% fewer queens than usual [51]. Neonicotinoid pesticides, even in little amounts, can lead to disorientation and memory impairment in forager bees [38], which in turn reduces the efficiency of pollen collecting by bumble bees [52]. Exposure to low dosages of the acaricide coumaphos also causes aberrant movement in honey bees, as shown in previous studies [53]. Without a doubt, all of these impacts disrupt the performance of both individual bees and the entire colony [54].

Ultimately, the consequential impacts brought about by herbicides must not be disregarded. Herbicides do not possess toxicity against bees, but they do disrupt the habitat of bees and other pollinators. Long-term use of herbicides has led to a significant decline in both plant biodiversity and the arthropod groups that depend on them [55, 56]. The absence of specific plant species, particularly weeds, indicates a decline in the quality of the natural habitat that supports pollinators, such as honey bees. As a result, bees face challenges in gathering the necessary assortment of pollen for a nutritious bee diet [57]. The shortage of blooms, which is caused by persistent pesticide treatments in agriculture and forestry regions for many decades, indirectly leads to poor bee nutrition.

4. RISK OF PESTICIDES TO BEES

To assess the true dangers that contemporary pest control agents and acaricides used for hive treatment represent to honey bees, it is necessary to consider the previously discussed routes of pesticide exposure and their specific effects on bees. The primary hazard arises from the high toxicity of the chemicals to bees, resulting in their death within a relatively short to medium timeframe. Additional hazards encompass sublethal consequences that can impair hive performance and jeopardize the long-term sustainability of bee colonies, as previously stated.

Risks are commonly assessed by calculating the likelihood of harm, taking into account the immediate harmful effects and the frequency at which a chemical can impact bees. There are three scenarios to consider: (1) the risks associated with spraying pesticides on agricultural fields; (2) the risks of consuming agrochemical residues found in pollen, honey, and water, which are collected and consumed by forager bees and then transferred to the hive, where they are turned into honey and beebread and fed to other bees, larvae, and the queen; and (3) the risks of being exposed to combs treated with acaricide products.

4.1. Risk from exposure to sprays

For the first scenario, the only data required are the concentrations of the active ingredients in the spray solutions applied and their acute toxicity, i.e., LD50 values for each chemical, since the probability of a bee being sprayed on can be considered 100% if the bee flies directly through the spray cloud in the field, or if a hive is placed downwind and within the normal range of spray drift by aerial or ground-rig applications, i.e., less than 1 km. This kind of risk is estimated using the typical hazard quotient HQ

$$HQ = \frac{Exposure(\mu g)}{LD50(\mu g / bee)} \qquad (1)$$

where the exposure term can be determined by the concentration of active ingredient in the spray droplets and the volume received by the bees, according to the following expression

$$HQ = \frac{Concentration(\mu g / ml) \times vol.droplets(ml)}{LD50(\mu g / bee)} \qquad (2)$$

In this case, the HQ can be indicative of high risk when its value is 1 or more, since 50% or more bees exposed would die; moderate risk is when HQ values are between 0.1 and 1 and low risk when it is less than 0.1, as fewer than 10% (similar to a natural mortality rate) of bees would be threaten.

Estimates of risks are typically done by considering the spray drift [58, 59] and the exposure to the flying bees [60]. For example, to compare the risk posed by different products under the same conditions, the spray drift volume can be fixed, e.g. 500 droplets for a bee crossing the spray cloud in a few minutes, at 5×10^{-4} µl for a standard droplet would result in 0.25 µl/bee. Table 2 shows a comparison of the risk that commonly applied pesticides would have in such situations.

Table 2. Risk of common agricultural pesticides to honey bees that fly across a spray cloud (ppm) and receive a total dose of 0.25 µl/bee

Pesticide type	Chemical name	Droplet concentration (µg/ml)	LD50 (µg/bee)	HQ	Risk evaluation
Acaricide	Amitraz	200	50.0	0.001	Low
	Dicofol	240	19.0	0.003	Low
	Propargite	600	62.1	0.002	Low
Fungicide	Azoxystrobin	75	200.0	<0.001	Negligible
	Fludioxonil	12.5	50.3	<0.001	Negligible
	Mancozeb	750	226.2	<0.001	Negligible
	Tolclofos-methyl	500	100.0	0.001	Low
Insecticide	Abamectin	18	0.03	0.15	Moderate
	Acetamiprid	225	7.9	0.007	Low
	Beta-cyfluthrin	25	0.031	0.20	Moderate
	Bifenthrin	100	0.015	1.70	High
	Carbaryl	500	0.84	0.15	Moderate
	Chlorantraniliprole	350	4.0	0.022	Low
	Chlorpyrifos	300	0.072	1.04	High
	Difenthiuron	500	1.5	0.083	Low
	Dimethoate	400	0.12	0.85	Moderate
	Endosulfan	350	6.35	0.014	Low
	Esfenvalerate	50	0.026	0.48	Moderate
	Fipronil	200	0.007	6.8	High
	Imidacloprid (spray)	200	0.061	0.81	Moderate
	Imidacloprid (dust)	24*	0.061	0.1	Moderate
	Indoxacarb	150	0.58	0.064	Low
	Lambda-cyhalothrin	250	0.048	1.3	High
	Methidathion	400	0.27	0.37	Moderate
	Methomyl	225	0.50	0.11	Moderate
	Pririmicarb	500	35.7	0.004	Low
	Spirotetramat	240	242	<0.001	Negligible
	Thiamethoxam (spray)	250	0.025	2.5	High
	Thiamethoxam (dust)	36.8*	0.025	0.37	Moderate

*Data for dust particles from conventional pneumatic planters [61, 62].

Table 2 demonstrates that the pesticides fipronil, thiamethoxam, bifenthrin, lambda-cyhalothrin, and chlorpyrifos pose the highest risk to bees when applied to agricultural crops. The microencapsulated version of lambda-cyhalothrin poses a significant risk due to the potential for bees to transport the concentrated chemical within the microcapsules to their hive. Typically, dust particles from seeds treated with neonicotinoids and droplets from sprays containing pyrethroids, organophosphorus, and carbamate insecticides have significant risks, ranging from moderate to high. On the other hand, other insecticides and acaricides show very low dangers in comparison. The fungicides depicted in this illustration, as well as potentially the majority of other fungicides administered as sprays on leaves, present minimal or insignificant hazards to bees through direct exposure to the droplets of the spray. This assessment aligns with the documented occurrences of pesticides affecting bees in both the United Kingdom [63] and Canada [64]. Undoubtedly, the insecticides with the highest toxicity levels pose the greatest threat to bees.

4.2. Risks by oral exposure

In the second scenario, it is crucial to have data on the concentration and frequency of residues in various sources, such as pollen, honey, and water. Additionally, it is important to get information on the nutritional intake of these sources by different groups of bees, including foragers, nurses, larvae, and the queen. The primary mode of exposure for bees in the hive is through ingesting contaminated food orally.

. The risk expression in this case would take the form

$$Risk = \frac{Frequency(\%) \times residue\,dose(\mu g)}{LD50(\mu g / bee)} \tag{3}$$

where the residue dose of a given pesticide can be estimated for different bees as the product of the concentration of residues in pollen, honey or water by the total intake of a particular caste of bee [21]. In turn, total intakes are estimated from daily intakes and the life span of bees, which vary from 5–6.5 days for larvae, 8–10 days for brood attendants and nurses, to 30 or more days for foragers [65]. The food intake by queens is hard to estimate, as they are fed royal jelly (a particular combination of pollen and honey), can live several years and vary their intake—which is unknown—throughout the reproductive and winter seasons. For the toxicity, oral LD50s are used in this case. The risk estimated by expression (3) can be interpreted as the probability that a given pesticide residue has of causing 50% mortality among the bees that ingest the contaminated pollen or nectar.

In recent years, a number of studies have reported the residue levels of agricultural pesticides found in pollen [66, 67] and nectar of flowers [68, 69], in water bodies of agricultural areas [28], as well as in beehive matrices, such as beebread, honey and wax [70, 71]. Based on these reports, we estimated the average and maximum residues for each pesticide as well as their frequency of appearance in those matrices. This information allowed us to calculate the risks that bees encounter when feeding on such contaminated food or drinking sources. A summary of results for the compounds that pose the highest risks by oral exposure of combined food and drink is shown in Table 3.

Table 3. Average pesticide residue levels in food and water (ppb) and their risk by oral exposure to worker honey bees and larvae. The time to reach the oral lethal dose (T50, days) is also shown for a comparison

Chemical name	Residues (μg/kg)			Larvae		Nurses		Foragers	
	Pollen	Honey	Water*	Risk (%)	T50 (days)	Risk (%)	T50 (days)	Risk (%)	T50 (days)
Thiamethoxam	28.9	6.4	4.1	2.77	23	4.80	27	276	7
Gamma-HCH (lindane)	7.6	176.5	-	0.62	9	0.01	979	200	3
Clothianidin	9.4	1.9	2.6	1.02	54	1.91	58	39.5	13
Imidacloprid (total)	19.7	6.0	0.9	1.19	68	1.57	103	25.4	25
Cypermethrin	13.9	18.1	-	0.13	119	0.04	711	4.00	44
Coumaphos (total)	128.3	105.5	-	0.11	1444	0.06	5524	2.62	545
Dinotefuran	45.3	13.7	-	0.10	49	0.13	74	1.50	20
Quinalphos	-	9.6	-	<0.01	253	-	-	1.29	91
Methiocarb	1.4	15.0	-	<0.01	1080	<0.01	>5000	1.08	391
Chlorpyrifos	32.6	3.9	-	0.04	1605	0.13	1118	0.86	764
Carbaryl	58.9	23.4	-	0.41	202	0.42	392	0.54	80
Beta-cyfluthrin	2.2	9.0	-	0.10	190	0.01	3497	0.43	69
Dimethoate	2.3	4.8	-	0.01	1198	<0.01	>5000	0.40	440
DDT (total)	31.2	44.2	-	<0.01	3871	<0.01	>5000	0.29	1432
Pirimiphos ethyl	-	19.0	-	<0.01	401	-	-	0.21	144
Diazinon	8.5	17.0	-	0.04	426	0.01	3869	0.19	156
Malathion	17.1	98.0	-	<0.01	3218	<0.01	>5000	0.15	1167
Pirimicarb	-	38.0	-	<0.01	3500	-	-	0.10	1261
Phosmet	339.3	-	-	0.07	991	0.79	168	-	-
Fipronil (total)	1.6	-	-	0.02	596	0.27	101	-	-
Acrinathrin (total)	146.8	-	-	0.01	719	0.17	122	-	-

*From the survey by Samson-Robert et al. [28].

Although neonicotinoids provide a significant risk, the majority of pesticide residues found in pollen and honey only pose a modest risk to bees, ranging from 1 to 5%. This is particularly true for pyrethroid and organophosphorus compounds.

Out of the 113 pesticide residues in food that have toxicity data, a total of 21 of them present a certain level of risk to honey bees. However, only 8% of these compounds are considered to be of concern. The concentration of residues in water varies significantly across different locations. The data presented in Table 3 is derived from a single study conducted in Canada, which only detected neonicotinoids, fungicides, and herbicides. However, the potential dangers associated with fungicides and herbicides were deemed to be insignificant, and therefore, they are not included in the table.

4.3. Risks by contact exposure

In addition to oral exposure, bee larvae can come into touch with residues that are deposited on the walls of the comb cells. These residues specifically include acaricides, which are used to treat Varroa. While the wax in a hive has the highest amounts of pesticide residues [23], it is believed that the presence of these compounds is generally limited, save for the acaricides used for fumigation. The potential harm to bee larvae from the later products should be assessed based on contact exposure rather than oral ingestion, as suggested by certain authors [30]. The expression (3) can be utilized, with the maximal residue dose in this instance evaluated at 5 mg of active chemical per cell for a solitary larva, substituting the contact LD50 for the oral LD50. Table 4 displays the outcomes of the risk analysis conducted on many acaricides in relation to honey bee larvae.

Table 4. Average acaricide residue levels in comb wax (ppb) and their risk to larvae of honey bees. The time to reach the lethal dose (T50, days) is also shown for a comparison

Pesticide	Residues in wax (µg/kg)	Risk by contact (%)	T50 (days)
Acrinathrin (total)	139.0	0.03	247
Amitraz (total)	585.5	<0.01	>5000
Bromopropylate	16.4	<0.01	>5000
Carbofuran (total)	19.4	<0.01	1649
Chlorfenvinphos	1156	0.14	709
Coumaphos (total)	1352	0.02	3003
Dicofol	6.8	<0.01	>5000
Pyridaben	5.4	<0.01	1957
Spirodiclofen	28.5	<0.01	>5000
Tau-fluvalinate	3144	0.15	551
Tau-fluvalinate+amitraz	3730	21.25	11
Tau-fluvalinate+coumaphos	4496	122.6	11
Tetradifon	7.9	<0.01	>5000

As it can be seen, the risks of acaricides to bee larvae are below 1% for all individual chemicals, but increases dramatically for synergistic mixtures, such as tau-fluvalinate with amitraz or coumaphos. Except for the latter mixtures, the overall risk to bee larvae of the individual products is very low or negligible compared to that of the same compounds by oral ingestion of contaminated food and water (Table 3).

4.4. Novel approaches to risk assessment

Another method for assessing hazards, specifically for oral exposures, involves determining the duration it would take for a bee to achieve the LD50 (lethal dose for 50% of the population) of a specific pesticide. This calculation is based on the daily consumption of contaminated food and water.

. This estimate is made using the expression

$$T50(days) = \frac{LD50(\mu g / bee)}{Daily\ \mathrm{intake}(\mu g)} \quad (4)$$

where T50 is the time to reach the median lethal dose (LD50), also termed median time to death. As it can be expected, there is a good correlation between the T50 values estimated using equation (4) and the risk values calculated using equation (3)—see Tables 3 and 4.

Neonicotinoid insecticides, however, can cause delayed mortality due to their agonistic mode of action [39]. This particularity means that their acute oral LD50s, which are usually estimated for exposures of 48 hours, are insufficient to estimate accurate risks of these insecticides, because the actual dose that causes the death of the bees decreases as the time of exposure increases [72]. Consequently, the mathematical function that relates the median time to death (T50) with the median lethal dose (LD50) is used to estimate the risk, as follows

$$Ln T50(days) = a + b Ln LD50(\mu g / bee) \quad (5)$$

where *a* (intercept) and *b* (slope) are empirical parameters specific to each chemical and species tested (in this case honey bees). The approach estimates the cumulative mortality with exposure time with more precision than the standard equation (4), as explained in a previous study [21].

4.5. Risk from Synergistic Mixtures of Pesticides

The aforementioned tables aid in identifying the pesticides that present the highest risk to bees, either by exposure to spray droplets or dust, consumption of contaminated food and water, or interaction with chemicals utilized for mite control within the hives. Evidently, the most majority (92%) of pesticides approved for agricultural use do not present substantial or quantifiable hazards to honey bees, but solely when taking into account the exposure to individual substances.

Recent findings suggest that the combination of specific chemicals, specifically insecticides and acaricides, with fungicides or mixes of acaricides, has a higher toxicity to bees compared to the individual compounds alone. The previous text has previously discussed the combined and enhanced effects of various combinations. To assess the dangers they provide, it is necessary to calculate them using the same methods, but adjusting the toxicity of the insecticide or acaricide based on a synergistic factor [21]. The values of these components are determined through experimental calculations for various combinations of fungicides with insecticides and/or acaricides [73]. Table 5 provides a few examples of these calculations.

Table 5. The synergistic effect of some fungicides with insecticides or acaricides and their risks to honey bees

Insecticide or acaricide	Fungicide	Synergistic factor	Risk to larvae (%)		Risk to nurses (%)	Risk to foragers (%)
			Wax	Food	Food	Food
Acetamiprid	Propiconazole	104.7	<0.01	<0.01	<0.01	<0.01
Acetamiprid	Fenbuconazole	4.5	<0.01	<0.01	<0.01	<0.01
Coumaphos	Fenpyroximate	20.0	<0.01	0.77	<0.01	<0.01
Cyhalothrin	Propiconazole	16.2	2.16	<0.01	<0.01	<0.01
Cyhalothrin	Myclobutanil	10.9	<0.01	<0.01	<0.01	<0.01
Cyhalothrin	Penconazole	4.4	<0.01	<0.01	<0.01	<0.01
Tau-fluvalinate	Myclobutanil	50.0	<0.01	0.01	<0.01	<0.01
Thiacloprid	Propiconazole	559.4	0.89	0.08	0.30	<0.01

While the overall increase in risk is evident, the specific combination of the pyrethroid insecticide cyhalothrin with propiconazole is the only one that poses a moderate worry for bee larvae. Even under these conditions, the danger associated with thiacloprid seems to be limited. Nevertheless, the larvae are exposed to a significant level of risk when specific combinations of acaricides, such as tau-fluvalinate with amitraz or coumaphos, are used in Varroa treatments. This information can be seen in Table 4.

5. MANAGEMENT IN ORDER TO AVOID PESTICIDE IMPACTS

The aforementioned risk assessments provide insights into the specific types of exposure that pose the most harm to distinct bee castes within the hives. Spray drift is the primary factor leading to incidents resulting in the death of forager worker bees [63, 74]. On the other hand, the consumption of contaminated pollen, nectar, and water is the underlying cause of Colony Collapse Disorder (CCD), which affects numerous apiaries worldwide [45]. CCD primarily affects nurse workers and, more specifically, the queen [49, 51]. Furthermore, the acaricides utilized in

Varroa treatment present a substantial hazard primarily to the bee larvae, and so jeopardize the long-term viability of the colonies. Recognizing these potential dangers can enable beekeepers and farmers to develop targeted strategies to mitigate the effects of pesticides.

Beekeepers must consider the surrounding landscape where their bees gather food, as a significant percentage of land in both developed and developing countries is dedicated to agriculture, where various types of pesticides are regularly employed. Given the essential role of these agrochemicals in agricultural production, it is not feasible to completely eliminate their use. Therefore, a logical strategy should focus on reducing the potential harm of these plant protection compounds to bees.

Chemical firms are legally required to indicate on product labels whether their goods pose a risk to bees. If this is the case, it is necessary for them to clearly state the potential hazards they present and provide explicit instructions, such as "DO NOT apply any spray to flowering plants while bees are actively collecting nectar." Nevertheless, the efficacy of label warnings is contingent upon the presence of adequate communication between applicators, farmers, and beekeepers. The former is obligated to notify beekeepers about any spraying activities, in order to relocate beehives to a secure area during the spraying season. Relocating beehives typically requires more than 24 hours, hence farmers must provide ample advance notice to their neighboring beekeepers. The only way to prevent damage to the hives is by avoiding drift.

Bees are typically active from sunrise to a couple of hours before dusk. Most honey bees gather food within a radius of 2-4 km from their hive, but they may travel up to 7 km or more when there is a shortage of pollen and nectar in their immediate vicinity [75]. Hence, the potential harm of pesticides to bees can be mitigated by applying them to crops during the late hours, when bees are not actively seeking food.

If a crop or weed-infested area has been accidentally sprayed, the farmer should inform the beekeepers who are affected so that they can respond accordingly. This measure will effectively prevent the managed bees from entering the sprayed area for a considerable period of time. In addition to affecting the cultivated areas, pesticides can also cause harm when they spread to the surrounding flora that is used by bees for foraging. This includes hedges, roadside weeds, and various types of trees such as fruit trees and eucalypts. Coolibah trees (Eucalyptus microtheca) thrive on flatlands adjacent to several river channels in the cotton cultivation regions of Australia. They serve as a key reservoir of nectar and pollen for both wild and honey bees. The Australian cotton industry has developed a document on best management practices that provides recommendations to cotton growers on how to address this issue and be mindful of the potential harm to beekeepers. The guideline states that by fostering effective communication and a positive attitude, apiarists and cotton growers may collaborate to mitigate hazards to bees. This is crucial considering the significance of both the honey industry and cotton sector in promoting regional development [33].

6. CONCISE SUMMARY

Overall, recognizing the negative effects of pesticides on bees should motivate all stakeholders to take necessary measures to reduce the chemical harm to bees and the productivity of the beekeeping business. The primary objective of such activities should be to effectively regulate the utilization of agrochemicals in a manner that does not adversely affect other land cultivators. Furthermore, producers should take measures to reduce the pollution of the surrounding environments, such as water bodies, with pesticides. This is important since not only honey bees, but also a wide range of pollinator species (such as butterflies, bumblebees, hoverflies, etc.) might be negatively impacted.

REFERENCES

1. vanEngelsdorp D, Meixner MD. A historical review of managed honey bee populations in Europe and the United States and the factors that may affect them. J Invertebr Pathol 2010; 103(Suppl): S80-S95.
2. EFSA. Towards an integrated environmental risk assessment of multiple stressors on bees: review of research projects in Europe, knowledge gaps and recommendations. EFSA J 2014; 12(3): 3594.
3. OECD Guidelines for the Testing of Chemicals: Honeybees, Acute Oral Toxicity Test. (1998).
4. Ellis J. The honey bee crisis. Outlooks Pest Manag 2012; 23: 35–40.
5. Potts SG, Roberts SPM, Dean R, Marris G, Brown MA, Jones R, et al. Declines of managed honey bees and beekeepers in Europe. J Apicult Res 2010; 49(1): 15–22.
6. Hald AB. Weed vegetation (wild flora) of long established organic versus conventional cereal fields in Denmark. Ann Appl Biol 1999; 134(3): 307–14.
7. Hyvonen T, Salonen J. Weed species diversity and community composition in cropping practices at two intensity levels: a six-year experiment . Plant Ecol 2002; 159(1): 73–81.
8. Goulson D, Nicholls E, Botías C, Rotheray EL. Bee declines driven by combined stress from parasites, pesticides, and lack of flowers. Science 2015; 347(6229): 1255957.

9. Johnson RM, Dahlgren L, Siegfried BD, Ellis MD. Acaricide, fungicide and drug interactions in honey bees (Apis mellifera). PLoS One 2013; 8(1): e54092.

10. Maini S, Medrzycki P, Porrini C. The puzzle of honey bee losses: a brief review. Bull Insectol 2010; 63(1): 153–60.

11. Krupke CH, Long EY. Intersections between neonicotinoid seed treatments and honey bees. Curr Opin Ins Sci 2015; 10: 8–13.

12. Craig I, Woods N, Dorr G. A simple guide to predicting aircraft spray drift. Crop Protection 1998; 17(6): 475–82.

13. Marinelli E, De Pace FM, Belligoli P, Oddo LP. Biomonitoring by bees. Tests in Latium region. Red ia 2004; 87: 207–10.

14. Greig-Smith PW, Thompson HM, Hardy AR, Bew MH, Findlay E, Stevenson JH. Incidents of poisoning of honeybees (Apis mellifera) by agricultural pesticides in Great Britain 1981–1991. Crop Protection 1994; 13: 567–81.

15. Georgiadis PT, Pistorius J, Heimbach U. Gone with the wind - drift of abrasive dust from seed treatments - a risk for honey bees (Apis mellifera L.)? [Vom Winde verweht - Abdrift von Beizstauben ein Risiko fur Honigbienen (Apis mellifera L.)?] Julius-Kuhn-Archiv 2010; (424): 33.

16. Bonmatin JM, Giorio C, Girolami V, Goulson D, Kreutzweiser DP, Krupke C, et al. Environmental fate and exposure; neonicotinoids and fipronil. Environ Sci Pollut Res 2015; 22(1): 35–67.

17. Krupke CH, Hunt GJ, Eitzer BD, Andino G, Given K. Multiple routes of pesticide exposure for honey bees living near agricultural fields. PLoS One 2012; 7(1): e29268.

18. Botias C, David A, Horwood J, Abdul-Sada A, Nicholls E, Hill EM, et al. Neonicotinoid residues in wildflowers, a potential route of chronic exposure for bees. Environ Sci Technol 2015; 49(21): 2731–12740.

19. Sánchez-Bayo F, Yamashita H, Osaka R, Yoneda M, Goka K. Ecological effects of imidacloprid on arthropod communities in and around a vegetable crop. J Environ Sci Health B 2007; 42(3): 279–86.

20. Tapparo A, Giorio C, Marzaro M, Marton D, Solda L, Girolami V. Rapid analysis of neonicotinoid insecticides in guttation drops of corn seedlings obtained from coated seeds. J Environ Monit 2011; 13(6): 1564–8.

21. Sánchez-Bayo F, Goka K. Pesticide residues and bees – A risk assessment. PLoS One 2014; 9(4): e94482.

22. Kubik M, Nowacki J, Pidek A, Warakomska Z, Michalczuk L, Goszczynski W. Pesticide residues in bee products collected from cherry trees protected

during blooming period with contact and systemic fungicides. Apidologie 1999; 30: 521–32.

23. Orantes-Bermejo FJ, Pajuelo AG, Megías MM, Fernández-Píñar CT. Pesticide residues in beeswax and beebread samples collected from honey bee colonies (Apis mellifera L.) in Spain. Possible implications for bee losses. J Apicult Res 2010; 48(1): 243–50.
24. Schmaranzer S. Thermoregulation of water collecting honey bees (Apis mellifera). J Ins Physiol 2000; 46(8): 1187–94.
25. Belden JB, Gilliom RJ, Martin JD, Lydy MJ. Relative toxicity and occurrence patterns of pesticide mixtures in streams draining agricultural watersheds dominated by corn and soybean production. Integr Environ Assess Manag 2007; 3(1): 90–100.
26. Siebers J, Binner R, Wittich K-P. Investigation on downwind short-range transport of pesticides after application in agricultural crops. Chemosphere 2003; 51(5): 397–407.
27. Woods N, Craig IP, Dorr G, Young B. Spray drift of pesticides arising from aerial application in cotton. J Environ Qual 2001; 30(3): 697–701.
28. Samson-Robert O, Labrie G, Chagnon M, Fournier V. Neonicotinoid-contaminated puddles of water represent a risk of intoxication for honey bees. PLoS One 2014; 9(12): e108443.
29. Martel A-C, Zeggane S, Aurieres Cm, Drajnudel P, Faucon J-P, Aubert M. Acaricide residues in honey and wax after treatment of honey bee colonies with ApivarR or AsuntolR50. Apidologie 2007; 38: 534–44.
30. Zhu W, Schmehl DR, Mullin CA, Frazier JL. Four common pesticides, their mixtures and a formulation solvent in the hive environment have high oral toxicity to honey bee larvae. PLoS One 2014; 9(1): e77547.
31. Chakrabarti P, Rana S, Sarkar S, Smith B, Basu P. Pesticide-induced oxidative stress in laboratory and field populations of native honey bees along intensive agricultural landscapes in two Eastern Indian states. Apidologie 2015; 46(1):107–29.
32. Zaluski R, Kadri SM, Alonso DP, Martins Ribolla PE, de Oliveira Orsi R. Fipronil promotes motor and behavioral changes in honey bees (Apis mellifera) and affects the development of colonies exposed to sublethal doses. Environ Toxicol Chem 2015; 34(5): 1062–9.
33. CottonInfo-Team. Cotton Pest Management Guide - 2013–2014. Australian Cotton Industry; Greenmount Press, Toowoomba (Qld) 2013. p. 164.
34. Boncristiani H, Underwood R, Schwarz R, Evans JD, Pettis J, vanEngelsdorp D. Direct effect of acaricides on pathogen loads and gene expression levels in honey bees Apis mellifera. J Ins Physiol 2012; 58(5): 613–20.

35. Sánchez-Bayo F. Impacts of agricultural pesticides on terrestrial ecosystems. In: Sánchez-Bayo F, van den Brink PJ, Mann R, editors. Ecological Impacts of Toxic Chemicals. Bentham Science Publishers; http://www.benthamdirect.com/51436/volume/1 2011. pp. 63–87.

36. Morrissey CA, Mineau P, Devries JH, Sánchez-Bayo F, Liess M, Cavallaro MC, et al. Neonicotinoid contamination of global surface waters and associated risk to aquatic invertebrates: a review. Environ Int 2015; 74: 291–303.

37. Hladik ML, Kolpin DW, Kuivila KM. Widespread occurrence of neonicotinoid insecticides in streams in a high corn and soybean producing region, USA. Environ Pollut 2014; 193: 189–96.

38. Decourtye A, Devillers J. Ecotoxicity of Neonicotinoid Insecticides to Bees. In: Thany SH, editor. Advances in Experimental Medicine and Biology - Insect Nicotinic Acetylcholine Receptors. Austin, TX: Landes Bioscience; 2009. pp. 85–95.

39. Rondeau G, Sánchez-Bayo F, Tennekes HA, Decourtye A, Ramírez-Romero R, Desneux N. Delayed and time-cumulative toxicity of imidacloprid in bees, ants and termite. Sci Rep 2014; 4: 5566.

40. Di Prisco G, Cavaliere V, Annoscia D, Varricchio P, Caprio E, Nazzi F, et al. Neonicotinoid clothianidin adversely affects insect immunity and promotes replication of a viral pathogen in honey bees. PNAS 2013; 110(46): 18466–71.

41. Vidau C, Diogon M, Aufauvre J, Fontbonne R, Viguès B, Brunet JL, et al. Exposure to sublethal doses of fipronil and thiacloprid highly increases mortality of honeybees previously infected by Nosema ceranae PLoS One 2011; 6(6): e21550.

42. Pettis J, vanEngelsdorp D, Johnson J, Dively G. Pesticide exposure in honey bees results in increased levels of the gut pathogen Nosema. The Science of Nature [Naturwissenschaften] 2012; 99(2): 153–8.

43. Doublet V, Labarussias M, de Miranda JR, Moritz RFA, Paxton RJ. Bees under stress: sublethal doses of a neonicotinoid pesticide and pathogens interact to elevate honey bee mortality across the life cycle. Environ Microbiol 2014; 17: 969–83.

44. Nazzi F, Brown SP, Annoscia D, Piccolo FD, Prisco GD, Varricchio P, et al. Synergistic parasite-pathogen interactions mediated by host immunity can drive the collapse of honeybee colonies PLoS Pathog 2012; 8(6): e1002735.

45. Sánchez-Bayo F, Goulson D, Pennacchio F, Nazzi F, Goka K, Desneux N. Are bee diseases linked to pesticides? — A brief review. Environ Int 2016; 89–90: 7–11.

46. Iwasa T, Motoyama N, Ambrose JT, Roe RM. Mechanism for the differential toxicity of neonicotinoid insecticides in the honey bee, Apis mellifera. Crop Protection 2004; 23(5): 371–8.

47. Pilling ED, Bromleychallenor KAC, Walker CH, Jepson PC. Mechanism of synergism between the pyrethroid insecticide λ-cyhalothrin and the imidazole fungicide prochloraz, in the honeybee (Apis mellifera L.). Pestic Biochem Physiol 1995; 51(1): 1–11.

48. DeGrandi-Hoffman G, Chen Y, Simonds R. The effects of pesticides on queen rearing and virus titers in honey bees (Apis mellifera L.). Insects 2013; 4(1): 71–89.

49. Sandrock C, Tanadini LG, Pettis JS, Biesmeijer JC, Potts SG, Neumann P. Sublethal neonicotinoid insecticide exposure reduces solitary bee reproductive success. Agric Forest Entomol 2014; 16(2): 119–28.

50. Sandrock C, Tanadini M, Tanadini LG, Fauser-Misslin A, Potts SG, Neumann P. Impact of chronic neonicotinoid exposure on honeybee colony performance and queen supersedure. PLoS One 2014; 9(8): e103592.

51. Whitehorn PR, O'Connor S, Wackers FL, Goulson D. Neonicotinoid pesticide reduces bumble bee colony growth and queen production. Science 2012; 336: 351–2.

52. Feltham H, Park K, Goulson D. Field realistic doses of pesticide imidacloprid reduce bumblebee pollen foraging efficiency. Ecotoxicology 2014; 23(3): 317–23.

53. Williamson SM, Wright GA. Exposure to multiple cholinergic pesticides impairs olfactory learning and memory in honeybees. J Exp Biol 2013; 216(10): 1799–807.

54. Desneux N, Decourtye A, Delpuech J-M. The sublethal effects of pesticides on beneficial arthropods. Annu Rev Entomol 2007; 52: 81–106.

55. Wardle D, Nicholson K, Bonner K, Yeates G. Effects of agricultural intensification on soil-associated arthropod population dynamics, community structure, diversity and temporal variability over a seven-year period Soil Biol Biochem 1999; 31(12): 1691–706.

56. Schmitz J, Hahn M, Brühl CA. Agrochemicals in field margins - An experimental field study to assess the impacts of pesticides and fertilizers on a natural plant community. Agric Ecosyst Environ 2014; 193: 60–9.

57. Wahl O, Ulm K. Influence of pollen feeding and physiological condition on pesticide sensitivity of the honey bee Apis mellifera carnica. Oecologia 1983; 59: 106–28.

58. Barmaz S, Potts SG, Vighi M. A novel method for assessing risks to pollinators from plant protection products using honeybees as a model species. Ecotoxicology 2010; 19: 1347–59.

59. Tasei JN. Impact of agrochemicals on non-Apis bees. In: Devillers J, Pham-Delegue MH, editors. Honey Bees: Estimating the Environmental Impact of Chemicals; CRC Press, London & New York 2002. pp. 101–31.

60. Poquet Y, Kairo G, Tchamitchian S, Brunet J-L, Belzunces LP. Wings as a new route of exposure to pesticides in the honey bee. Environ Toxicol Chem 2015; 34(9): 1983–8.

61. Tremolada P, Mazzoleni M, Saliu F, Colombo M, Vighi M. Field trial for evaluating the effects on honeybees of corn sown using cruiser® and Celest® treated seeds. Bull Environ Contam Toxicol 2010; 85(3): 229–34.

62. Pochi D, Biocca M, Fanigliulo R, Pulcini P, Conte E. Potential exposure of bees, Apis mellifera L., to particulate matter and pesticides derived from seed dressing during maize sowing. Bull Environ Contam Toxicol 2012; 89(2): 354–61.

63. Barnett EA, Charlton AJ, Fletcher MR. Incidents of bee poisoning with pesticides in the United Kingdom, 1994–2003. Pest Manag Sci 2007; 63(11): 1051–7.

64. Cutler GC, Scott-Dupree CD, Drexler DM. Honey bees, neonicotinoids and bee incident reports: the Canadian situation. Pest Manag Sci 2014; 70(5): 779–83.

65. Rortais A, Arnold G, Halm MP, Touffet-Briens F. Modes of honeybees exposure to systemic insecticides: estimated amounts of contaminated pollen and nectar consumed by different categories of bees. Apidologie 2005; 36(1): 71–83.

66. Skerl MIS, Bolta SV, Cesnik HB, Gregorc A. Residues of pesticides in honeybee (Apis mellifera carnica) bee bread and in pollen loads from treated apple orchards. Bull Environ Contam Toxicol 2009; 83(3): 374–7.

67. Kasiotis KM, Anagnostopoulos C, Anastasiadou P, Machera K. Pesticide residues in honeybees, honey and bee pollen by LC-MS/MS screening: reported death incidents in honeybees. Sci Total Environ 2014; 485-486(0): 633–42.

68. Wallner K. Sprayed and seed dressed pesticides in pollen, nectar and honey of oilseed rape. Julius-Kuhn-Archiv 2009; (423): 152–3.

69. Byrne FJ, Visscher PK, Leimkuehler B, Fischer D, Grafton-Cardwell EE, Morse JG. Determination of exposure levels of honey bees foraging on flowers of mature citrus trees previously treated with imidacloprid. Pest Manag Sci 2014; 70(3): 470–82.

70. Mullin CA, Frazier M, Frazier JL, Ashcraft S, Simonds R, vanEngelsdorp D, et al. High levels of miticides and agrochemicals in North American apiaries: implications for honey bee health. PLoS One 2010; 5(3): e9754.

71. Chauzat M-P, Faucon J-P, Martel A-C, Lachaize J, Cougoule N, Aubert M. A survey of pesticide residues in pollen loads collected by honey bees in France. J Econ Entomol 2006; 99(2): 253–62.

72. Tennekes HA, Sánchez-Bayo F. The molecular basis of simple relationships between exposure concentration and toxic effects with time. Toxicology 2013; 309: 39–51.

73. Boobis A, Budinsky R, Collie S, Crofton K, Embry M, Felter S, et al. Critical analysis of literature on low-dose synergy for use in screening chemical mixtures for risk assessment. Crit Rev Toxicol 2011; 41(5): 369–83.

74. Mineau P, Harding KM, Whiteside M, Fletcher MR, Garthwaite D, Knopper LD. Using reports of bee mortality in the field to calibrate laboratory-derived pesticide risk indices. Environ Entomol 2008; 37(2): 546–54.

75. Beekman M, Ratnieks FLW. Long-range foraging by the honey-bee, Apis mellifera L. Funct Ecol 2000; 14(4): 490–6.

05

PROGRESS IN PHARMACOLOGICAL EFFECTS AND CHEMICAL COMPOSITION OF PROPOLIS MANUFACTURED IN THE AMERICAS

1. INTRODUCTION: AN OVERVIEW

Bees are the primary pollinators for blooming plants, playing a crucial role in their ecological balance. This coevolutionary relationship between bees and plants has been ongoing for over 100 million years. Eusocial bees have achieved evolutionary success by living in perennial colonies consisting of approximately 50,000 individuals. They have also developed advanced recruitment communication mechanisms to forage and exploit the chemical properties of plants. Additionally, they manufacture and apply beehive products for their own advantage [1–5]. Bees have gained a key place in various civilizations throughout history and different geographical regions due to their crucial role as carriers of pollen in agricultural crops and the impact of their hive products on human societies. The technique of managing bees has been documented from ancient times, encompassing two distinct types of beekeeping: apiculture, which focuses on the Apinae species, and meliponiculture, which focuses on the Meliponinae species. The first category consists of the Asian honeybees (Apis cerana) and western European honeybees (Apis mellifera), while the second category pertains to the indigenous tropical and subtropical stingless bees (Melipona sp., Oxytrigona sp., Scaptotrigona sp., Tetragonisca sp., and Trigona sp., among others). Currently, the practice of beekeeping with the western European honeybee, A. mellifera, is geographically extensive due to human migrations. As a result, A. mellifera has established itself in all the areas where humans have settled, covering a wide range of phytogeographical distribution, including nearly every vegetated location on the planet [2, 3, 6–10].

The eusocial bees in the hive demonstrate cooperative behavior by each individual contributing to the maintenance of the colony. This results in a healthy community known as social immunity, which is characterized by hygienic practices. These practices include removing diseased brood to prevent the spread of pathogens and parasites within the hive. Nevertheless, a costly outcome of living in social groups is the transmission of diseases caused by frequent interactions between individuals. To provide extra immunological advantages, bees gather antimicrobial natural resins and use them to create a substance known as propolis [4, 11–13]. Propolis, including honey, beeswax, and royal jelly, is a valuable product derived from beehives that has been highly esteemed by human communities throughout history. Honeybees (A. mellifera) typically create propolis by selectively collecting resins from various plant sources such as leaves, buds, sap flows, and trichomes. These resins are then combined with beeswax and secretions from the salivary glands, resulting in a chemically intricate resinous substance [1, 14–18]. The collected secretions are primarily integrated into propolis without undergoing any chemical alterations. However, some glycosides are exposed to enzymatic activity by salivary hydrolases produced by bees [19, 20]. Stingless bees create a unique substance called geopropolis by combining soil and clay particles with plant exudates and beeswax. This substance is different from the propolis produced by honeybees because it contains minerals and lacks plant trichomes [6, 20-22].

Propolis varieties produced by honeybees and stingless bees have both mechanical and biological properties. These materials are used within the hive to seal cracks, prevent structural damage, and serve as a thermoregulatory resource. Additionally, these products act as chemical weapons to protect the colony from diseases, functioning as antimicrobial agents and embalming substances that prevent the putrefaction of killed intruders [1, 14–18, 23, 24]. Furthermore, propolis has been utilized as a therapeutic treatment in various human traditional medicine systems, owing to its inherent qualities. Indeed, the Ancient Egyptians, Greeks, and Romans primarily utilized this adhesive substance for the purpose of wound healing and as an agent with antibacterial properties. Furthermore, throughout Central and South America, the Maya and Inca civilizations utilized cerumen and geopropolis created by stingless bees as a traditional cure [15, 20, 25–28]. In contemporary times, propolis is utilized in alternative medicine in Japan, specifically as a treatment for wounds, burns, sore throat, and stomach ulcers. Similarly, geopropolis is employed by the inhabitants of certain tropical countries in the Americas as an empirical remedy for wound healing, gastritis, infections, and other ailments. Propolis and geopropolis are considered to be a highly promising source of bioactive chemicals for pharmacological study, as indicated by multiple studies [1, 14–18, 21, 28, 29].

Extensive research has been conducted on propolis, and in recent decades, propolis has garnered scientific interest. Numerous publications have been published on its wide range of pharmacological activity and its bioactive components

[14, 15, 18, 28]. The current biological activities attributed to propolis include antibacterial, antioxidant, antiparasitic, antifungal, antiviral, local anaesthetic, anti-inflammatory, immunomodulatory, antitumor, and antiproliferative effects on cancer cell growth, among others. The biological activities of propolis vary significantly depending on its geographical origin. These changes are primarily determined by the qualitative and quantitative differences in its unique chemical ingredients, which are derived from different botanical sources. Therefore, the chemical variety of propolis is determined by the phytogeographical conditions and climatic parameters, as well as the specific honeybee species responsible for its synthesis [55–57]. Propolis from temperate zones (Europe, North America, Southern South America, and West Asia) has a different chemical makeup compared to propolis from tropical zones (Central and South America, South and Southern Asia, and Africa), as indicated by numerous recent research.

The primary botanical sources of propolis in temperate regions include exudates from poplar buds (Populus spp.), as well as birch (Betula alba L.), horse chestnut (Aesculus hippocastanum L.), alder (Alnus glutinosa Medik), beech (Fagus sylvatica L.), and certain conifers. Currently, more than 300 chemical components have been discovered in various temperate propolis. These chemicals include phenolic acids and esters, flavonoids, terpenes, lignans, aromatic aldehydes and alcohols, fatty acids, stilbenes, and steroids [14, 15, 18, 58]. In propolis samples from various tropical regions, including Baccharis dracunculifolia, Araucaria angustifolia, Clusia minor, Clusia rosea, Dalbergia ecastophyllum, Macaranga tanarius, Hyptis divaricata, and Eucalyptus citriodora, among others, more than 250 compounds have been identified. These compounds include prenylated benzophenones, organic acids, prenylated organic acids, terpenes, alcohols, and isoflavonoids [58, 59]. However, it has been noted that even while propolis generated by A. Mellifera comes from plants, its total composition remains rather constant. It consists of 50% botanical resins, 30% waxes, 10% essential and aromatic oils, 5% pollen, and 5% miscellaneous organic compounds. Due to the chemical diversity found in propolis, its sensory qualities can vary significantly. This includes its physical appearance, texture, smell, and color, which can range from dark-brown to yellow [22, 58, 62–64].

In order to conduct pharmacological and chemical experiments, it is necessary to purify propolis. The elimination of inert substances, primarily wax, is typically accomplished by creating alcoholic or hydro-alcoholic extracts of the macerated raw material, commonly employing ethanol, 70% ethanol, and methanol. Various analytical methods, such as thin-layer chromatography (TLC), gas chromatography (GC), high-performance liquid chromatography (HPLC), and their combination with mass spectrometry (MS), have been employed to identify and describe the chemical components of propolis samples from different geographical locations

[1, 58]. GC–MS analysis has been employed to identify the volatile and semi-volatile constituents of propolis. However, certain components of propolis are not enough volatile for GC–MS analysis, even when derivatization processes are utilized. Electrospray ionization (ESI) has been widely employed to ionize nonvolatile, thermally unstable, heavy, and polar compounds. As a result, ESI-MS and its tandem version ESI-MS/MS have been utilized for the analysis of propolis. Electrospray ionization (ESI) is more efficient at ionizing polar compounds that contain acid groups in negative ion mode (-) or basic sites in positive ion mode (+). This is an important feature for studying the chemistry of propolis, as many of its pharmacologically active components have a certain level of polarity, often due to the presence of acidic phenolic groups [65, 66]. Furthermore, nuclear magnetic resonance (NMR) techniques have been employed to thoroughly analyze and describe the isolated chemicals derived from propolis [1, 58].

In recent years, researchers have examined the chemical composition and pharmacological effects of propolis samples collected from various regions in North America, including Central America and the Caribbean, as well as South America. Numerous scientific papers have documented both the differences and similarities observed among the analyzed samples from different geographical locations. Currently, propolis from the Americas is categorized based on its appearance and the plant it comes from. This is because the continent's biological diversity results in many types of botanical resins. The main categories of propolis found in the Americas are the North and Southern South American propolis, referred to as the 'poplar' type, as well as the Cuban, Brazilian, and Mexican 'red' propolis type. Additionally, there is the Brazilian "green" type, and finally the Cuban and Venezuelan "Clusia" type. These different types of propolis are a result of the European honeybees adapting to the specific needs of their beehives and the vegetation found in the Americas, both in tropical and temperate regions. Furthermore, the propolis created by stingless bees in the Americas has attracted attention, and several samples have undergone chemical and biological analysis in recent years. In certain instances, these samples have shown parallels in both bioactivity and composition to the propolis generated by honeybees in tropical regions. Nevertheless, in this chapter, our sole emphasis was on honeybee propolis.

Several authors have conducted a thorough examination of reports on propolis research using various methods [1, 14, 15, 18, 21, 28, 55, 58, 62]. However, there has been no discussion on the analysis of propolis from the Americas in an effort to compare its bioactivity, phytochemical diversity, and botanical sources. In this paper, we provide a comprehensive overview and analysis of the existing research on honeybee propolis in the Americas. We also highlight the latest advancements in the study of this valuable product for beekeeping. Figure 1 illustrates the majority of studies conducted on propolis from the Americas.

Figure 1. Chemical and pharmacological studies on propolis produced by Apis mellifera on Americas. The black dots on the Map indicate the geographical origin of studied propolis.

2. NORTH AMERICAN PROPOLIS

North America encompasses the geographic area that includes Greenland, Canada, the United States, and Mexico, as well as the countries within Central America and the Caribbean Islands. North America is a terrestrial part of the Americas continent, known for its abundant biodiversity and vast reserves of natural resources that are spread across its landmass. Furthermore, this landmass exhibits many topographical features, including valleys and huge mountain ranges. It is also known for its wide array of temperatures and biomes, including as tundra, subarctic forest, temperate forest, plains, deserts, and tropical forests [67, 68].

Despite not being indigenous to North and South America, the honeybee is extensively scattered over the Americas. A. mellifera was introduced to the Western Hemisphere by European colonists in the early seventeenth century due to its adaptability and its potential for human use [9, 69]. Subsequently, scientists introduced the African honeybee (A. mellifera scutellata) into Southeastern Brazil in the mid-1950s. Due to the accidental escape of African queen bees, there has been a gradual process of Africanization of both feral and managed A. mellifera colonies in the tropical and subtropical regions of North and South America [70–72]. Before and after the process of Africanization, propolis gathered from Canada, United States, Mexico, Cuba, Honduras, and El Salvador has undergone chemical and pharmacological analysis. In this summary, we present the key findings.

2.1. Canada

Canadian propolis has been subjected to chemical analysis using chromatographic techniques, specifically GC–MS and HPLC–MS. This analysis has successfully detected the presence of chemicals that are distinctive of poplar. As far as we know, a study on the chemical makeup of Canadian propolis was conducted using a sample obtained from the Sydenham region in the state of Ontario [73]. A total of fifteen flavonoids were identified, which included pinobanksin-3-O-alkanoates and methyl ether derivatives of quercetin. The primary components found in this Canadian propolis were the flavanones pinocembrin and pinobanksin-3-O-acetate, as well as the flavones chrysin and galangin. The presence of significant quantities of these compounds indicated that the propolis originated from American poplars P. deltoides, P. fremontii, and P. maximowiczii, which belong to the Aigeiros section and are known for their ability to produce these compounds [73].

Poplars from the Tacamahaca and Leuce sections are equally widespread in North America as poplars from the Aigeiros section. In order to ascertain the chemical composition of propolis from Canada that was collected outside the geographic zone where Aigeiros poplars are typically found, two samples from distinct regions, namely the Boreal forest in Richmond, Québec, and the Pacific coastal forest in Victoria, British Columbia, were subjected to analysis using GC-MS. The primary component found in Victorian propolis was p-hydroxyacetophenone, with benzyl hydroxybenzoate, cinnamic acid, and substantial quantities of five dihydrochalcones also present. The identification of these chemicals indicates that the plant source of propolis from the Canadian Pacific coast is likely to be a specific type of poplar known as P. trichocarpa Torr. et Gray, which belongs to the Tacamahaca section [74]. The primary components identified in propolis from the Richmond region were p-coumaric and cinnamic acids and their derivatives. The abundance of these compounds, along with the low concentration of flavonoids, is a distinctive feature of resins from Populus spp. in section Leuce. This suggests that P. tremuloides, a common type of poplar found in the Canadian Boreal forest, is likely the botanical source of propolis in Richmond. The DPPH tests were used to determine the antioxidant or antiradical activity of Canadian propolis collected from British Columbia and Québec. Both samples exhibited a strong ability to scavenge free radicals, with an activity of around 26 μg/mL resulting in a scavenging rate of 79 ± 5% and 65 ± 7%, respectively. These results were attributed to the presence of various phenolic compounds reported earlier [74].

Additional research has been conducted on the chemical makeup of commercially available extract formulations of Canadian propolis, specifically Herstat® propolis extract ACF®. This extract, which is patented, is produced using propolis collected from a specific region in Western Canada, likely Manitoba, known for its abundance of poplar trees. The extract contains a 3% concentration of propolis and is known for its antiviral properties due to the presence of flavonoids. Bankova et al. [75] discovered through GC-MS analysis that the primary components of the ointment formulation being sold were benzoic acid, p-coumaric acid, benzyl p-coumarate, a group of dihydrochalcones, as well as pinocembrin chalcone and pinostrobin chalcone. Significantly, the existence of these chemicals indicates a combination of botanical sources, specifically the exudates of both P. balsamifera and P. tremuloides [75]. Clinical studies were conducted to determine the antiviral activity of Herstat® propolis extract against both types of Herpes simplex infection: HSV-1 and HSV-2. The results showed that this ointment was more effective than acyclovir treatment in reducing healing time and alleviating local symptoms of cold sores [76, 77]. In order to comprehend the underlying antiviral properties of Herstat® Canadian propolis extract, Bankova et al. [75] conducted in vitro tests to determine its virucidal effect on HSV-1 and HSV-2, as well as its ability to prevent the adsorption of HSV-1 virus on MDBK bovine kidney cells. The in vitro findings aligned with the results from organized clinical trials including the topical ointment, providing evidence for the efficacy of propolis extract in treating herpes virus lesions [75].

In addition, a different commercial Canadian propolis obtained from TrophicTM products was examined in order to identify its composition and evaluate its antioxidant qualities. The commercial propolis was extracted utilizing a two-step sequential procedure involving ethanol and water. The ethanolic extracts showed a stronger antioxidant activity, as determined by the FRAP and DPPH techniques, as well as a higher quantity of polyphenols and flavonoids. In addition, the ESI-MS fingerprints showed similarities with Brazilian brown propolis and the presence of chrysin and pinocembrin, along with other flavonoids [78]. These studies on Canadian propolis provide sufficient phytochemical evidence regarding the botanical source of these bioactive resins, which are primarily collected from various poplar species throughout the country. This confirms that the chemical composition of the propolis is characteristic of temperate zones.

2.2. United States

The introduction of A. mellifera to the United States from England is documented to have occurred around 1622, first in the Colony of Virginia and thereafter in other

Colonies in the Eastern part of North America [69]. Recently, numerous studies have focused on propolis collected in various regions of the United States. One of the initial studies aimed to assess the inhibitory effects of propolis (collected in Illinois) against Paenibacillus larvae ssp. larvae, the causative agent of American foulbrood. A solution of propolis had a toxic impact on P. larvae in laboratory conditions (at a concentration of 10 μg/mL). However, further investigations conducted in living organisms indicated that propolis, when provided to infected colonies, only had a temporary inhibitory effect on bacterial growth, allowing P. larvae to multiply in the hive following treatment [79, 80].

As far as we know, the further investigations conducted on propolis from the United States were primarily concerned with identifying the main chemical components of samples from Western Ohio, as well as North and South Georgia (specifically Athens and Claxton). In these samples, similarities to the flavonoid profile of European propolis were observed in Ohio and North Georgia, along with the presence of kaempferol, galangin, 3,3′-dimethoxyquercetin, and 3-methoxykaempferol in Ohioan propolis [81]. The flavonoids are naturally present in some species of Populus, Pinus, Betula, Alnus, and Aesculus (horse chestnut). These plant sources are visited by honeybees for resins [16] and have been suggested by the authors as potential botanical origins of Ohioan propolis. Curiously, the Ohioan sample of propolis did not include any of the aromatic acids (ferulic, caffeic, and cinnamic acid) that are typically found in European propolis. This distinction distinguishes the Ohioan propolis from the European kind [81].

In a comparative study conducted by Kumazawa et al. [82], using various geographical samples from around the world, it was determined through HPLC analyses that a propolis sample from the United States (supplied by Api corporation and Tamagawa University, Japan) primarily consisted of aromatic acids, flavonoids, and their ester and methyl ether derivatives. The presence of various compounds such as p-coumaric acid, pinobanksin-5-methyl ether, pinobanksin, chrysin, pinocembrin, galangin, pinobanksin-3-O-acetate, and tectochrysin indicates that the exudates from P. tremuloides and poplars from section Aigeiros (Central and Eastern America), specifically P. fremontii, P. deltoides, or P. maximowiczii, have a mixed botanical origin [18, 74, 75, 82]. Furthermore, the propolis from the United States exhibited a moderate FRS (Free Radical Scavenging) activity. This was determined through the DPPH assay, where it showed a minimum activity of 50% at a concentration of 20.0 μg/mL. Additionally, it demonstrated a minimum activity of 30% at a concentration of 10.0 μg/mL in the β-carotene–linoleic acid system. In comparison, propolis samples from China, Australia, New Zealand, and Hungary exhibited higher FRS activities, with minimum activities of 70% at 20.0 μg/mL and 60% at 10.0 μg/mL, respectively [82]. The study aimed to compare the antimicrobial effectiveness of propolis from the United States, Turkey, and

Australia against oral pathogen microorganisms (P. gingivalis, P. intermedia, C. rectus, F. nucleatum, C. albicans, C. parapsilosis, C. krusei) without causing any harm to human gingival fibroblasts. Both United States and Australian propolis shown significant growth-inhibitory action against the studied microorganisms. However, it was noted that at the same quantities (dilution of 1:256 for bacteria and 1:2048 for Candida species), there was a cytotoxic effect on gingival fibroblasts [83].

A comparative examination was conducted on propolis samples from several regions, including the Eastern United States (Indiana and New York), Europe (Bulgaria, Finland, and England), and Brazil. The analysis method used was ESI-MS/MS. Chrysin, pinocembrin, and p-coumaric acid were detected in all of the propolis samples, indicating a shared botanical source in poplar species found in Eastern United States, South Brazil, and Europe [84]. In addition, propolis samples collected from Oregon and Northern California (three samples from each location) were analyzed using GC-MS and HPLC techniques. These analyses revealed that these samples contained a high concentration of p-hydroxyacetophenone, p-coumaric acid, t-cinnamic acid, and flavonoids such as galangin, chrysin, pinocembrinin, and pinobanksin derivatives. Furthermore, terpenes, terpenoids, chalcones, and dihydrochalcones were also present in these samples. These chemical compounds are typically found in the resin of balsam poplars belonging to the Tacamahaca section in the western part of the North American continent [85]. It is worth noting that ferulic acid and caffeic acid esters were found in samples from Oregon, but only in one specimen from California. These chemicals are usually found in resins from poplars of the Aigeiros section (cottonwoods) and are not present in resins from poplars of the Tacamahaca section [86]. Collectively, these findings suggest that two samples from California are exclusively derived from Tacamahaca poplars, while the other propolis specimen from California and the three samples from Oregon have a combination of plant origins from both Tacamahaca and Aigeiros poplars. This is due to the presence of caffeic acid esters and ferulic acid [85].

A recent study conducted a thorough analysis using gas chromatography-mass spectrometry (GC-MS) on ten different propolis samples collected from various regions in the United States, including the cold North, the wet Southeast, and the dry Southwest. The study utilized a chemometric approach called principal component analysis (PCA) to classify the samples based on the relative amounts of the main chemical classes present in each sample. An analysis was conducted on propolis samples collected from various locations in the United States, including New York (NY-2,3,8,10), Pennsylvania (PN-6), Louisiana (LA-1), Minnesota (MN-9), Nebraska (NE-4), Nevada (NV-5), and North Carolina (NC-7). The analysis revealed the presence of more than 60 chemical constituents, which were categorized into different compound types such as benzoic and cinnamic acid derivatives, chalcones, flavanones and dihydroflavonols, flavones and flavonols, phenolic glycerides, and terpenes. This information was obtained from a study cited as reference [87]. As anticipated, several geographical samples exhibited unique chemical signatures. Poplar-type propolis components, including aromatic acids and their esters, flavonoids, and chalcones, were detected in all samples. Three primary categories were acquired: The samples NY-2, NY-3, MN-9, and NY-10 belong to a group of propolis that is abundant in cinnamic acid derivatives. These derivatives include benzoic acid, cinnamic acid, p-coumaric acid, ferulic acid, and benzyl-p-coumarate. This particular type of propolis is found in the exudates of P. tremuloides Michx., which is commonly known as American aspen. The American aspen is the botanical source of these samples [87].

Group II) Propolis containing elevated levels of flavonoids (NE-4, PA-6, and NY-8), including pinocembrin, pinobanksin, pinobanksin-3-O-acetate, chrysin, galangin, and pinocembrin chalcone. This chemical composition is characteristic of poplar bud resins found in section Aigeiros. Resins from P. fremontii were identified as the primary source. Group III) This type of propolis contains high levels of triterpenes, specifically LA-1, NV-5, and NC-7. One specific triterpenic acid, called 3-oxo-6β-hydroxy-lup-20(29)-en-28-oic acid, has been found in Honduran propolis. This suggests that other botanical resins, particularly from the Liquidambar styraciflua L. tree (the main source of Honduran propolis), contribute to the propolis. This tree is commonly found in the eastern and southeastern regions of the United States and is known to contain this triterpene, as well as benzyl p-coumarate and cinnamic acid derivatives. The discovery of triterpenes in North American propolis is significant because previously, triterpenes had only been found in propolis samples from tropical and subtropical climates [58, 87, 88]. Furthermore, the researchers assessed the quorum sensing inhibitory (QSI) activity of ten propolises from the United States. They used the acyl-homoserine lactone (AHL)-dependent Chromobacterium violaceum strain CV026 to see if these propolises had the ability to inhibit the communication process in bacteria. The objective was to uncover any potential antivirulence properties in these samples. Group II demonstrated the most pronounced QSI (quorum sensing inhibition) effect. This classification offered a new perspective on the composite botanical source of certain propolis found in the United States, in addition to the previously identified poplar-type propolis [87].

A separate study was conducted to analyze the antimicrobial properties of propolis against pathogens found in beehives. A total of 12 samples were collected from various locations across the United States, including Chaska, MN; Baton Rouge, LA; Ithaca, NY; Jamestown, ND; Lincoln, NE; Raleigh, NC; Wakinsville, GA; Tucson, AZ; Aspen, CO; Vacaville, CA; Beaumont, TX; and Fallon, NV. These samples were then tested against two bee pathogens, P. larvae and Ascosphaera apis. An LC-MS-based metabolomic analysis was conducted to examine the chemical composition of propolis samples. This analysis revealed variations in chemical patterns and the varying ability of propolis samples to hinder the growth of infections. The propolis from Nevada showed the maximum activity against P. larvae and A. apis, with IC50 values of 41.6 and 8.6 μg/mL, respectively. The propolis from Texas had somewhat lower activity, with IC50 values of 46.9 and 10.0 μg/mL, respectively. The propolis from California had the lowest activity, with IC50 values of 74.1 and 7.4 μg/mL, respectively [13].

To determine the botanical source of antimicrobial resins collected by A. mellifera bees in Northeastern United States, we conducted an investigation using HPLC and UPLC-TOF-MS on plant material obtained from individual honeybees in a Minnesota apiary. Subsequently, the phytochemical patterns of resinous material taken from 6 North American Populus spp. and 5 hybrids, as well as other plants in the surrounding areas, were studied and compared using metabolomic methods, namely principal component analysis. The findings indicated that honeybees only collected resins from P. deltoides and P. balsamifera, disregarding numerous other available plant sources, even those that were chemotaxonomically similar. Out of the 26 individual resin foraging bees, 10 bees collected resin from P. deltoides and 15 bees collected resin from P. balsamifera. Furthermore, geographical or seasonal influences did not demonstrate any influence on the phytochemicals found in P. deltoides and P. balsamifera resins. The research indicated that honeybees can differentiate between closely related Populus species in order to gather resinous material. This behavior is in addition to the foraging behavior exhibited by individual bees within the hive, which is limited to a single plant source. Ultimately, the antibacterial impact of Populus spp. resins on P. larvae exhibited varying levels of inhibition due to differences in secondary metabolites found in the resins [4].

Additional chemical analyses have been conducted on propolis gathered from the Sonoran Desert, which is located in the Southwestern United States and falls outside the temperate poplar zone of North America. Wollenweber and Buchmann [89] conducted an analysis of many propolis samples obtained from both managed and feral honeybee colonies in Arizona State. Their objective was to identify the specific plants that the propolis in arid areas originated from. The vegetation in the Sonoran Desert is predominantly composed of xeromorphic shrubs and cactus, with poplars being rare and mostly limited to certain areas near watercourses. Wollenweber and Buchmann used TLC and GC-MS to identify that certain propolis samples had the unique pattern of P. fremontii. Additionally, some samples showed a combination of P. fremontii and Ambrosia deltoidea (Torrey) Payne. On the other hand, propolis samples that were not within the flight range of poplars contained flavonoid and other phenolic compounds that are specific to plants in that area, such as Ambrosia deltoidea and Encelia farinosa A. Gray [89].

The Hawaiian Islands chain is part of the political territory of the United States, in addition to the mainland. It is characterized by a distinct biological scenario with tropical features. Poplars are not accessible for propolis production in Hawaii. Therefore, honeybees need to rely on alternative botanical sources. In a recent study, Inui et al. [90] conducted a chemical analysis of Hawaiian propolis using HPLC–ESI-MS to determine its botanical source. This propolis sample contained a group of prenylflavonoids, which are also found in Pacific propolis types such as Okinawa, Taiwan, Indonesia, and Myanmar propolis. Macaranga tanarius was shown to be the primary source of the chemical profile, according to a comparative analysis [90]. Honeybees have been shown to collect exudates from the buds and bark of Plumeria acuminata, P. acutifolia, Schinus terebinthifolius, and Psidium guajava, as indicated by previous research [18].

The investigations conducted on United States propolis have shown that A. mellifera mostly visits Populus species to collect their valuable antibacterial resins. The phytochemical evidence confirms that various complementary plants produce appealing resins for honeybees and serve as the primary chemical source of propolis collected from different climatic regions in the continental United States. These plants include L. styraciflua, A. deltoidea, E. farinosa, and M. tanarius in the tropical Hawaiian islands. In addition, Wilson et al. [4] have made a significant contribution by highlighting the foraging fidelity of a solitary individual within the colony, exclusively collecting resins from a single plant source. This finding suggests that individuals within the colony exhibit organization and specialization towards specific plants, thereby contributing to the chemical diversity within the hive. These investigations highlight the diverse range of propolis types found in different ecological zones of the United States. They demonstrate that specific samples have mixed botanical origins, resulting in a wide range of pharmacological actions.

2.3. Mexico

Mexico is classified as one of the five countries in the world that has a significant number of unique species. This distinction is primarily due to the country's extensive range of different landforms and the diverse climate zones that span from North American deserts to Mesoamerican forests [91]. Analyzed propolis samples from several ecological zones in Mexico include those from the Sonoran Desert in North America, tropical forests, southern semi-arid highlands, and temperate sierras.

Sonoran Desert propolis is a well researched form of propolis from Mexico. For nearly a decade, numerous studies have documented the biological activity and primary chemical composition of samples obtained from dry and semi-arid regions in the Sonora State. These lands include Ures (UP), Pueblo de Alamos (PAP), and Caborca (CP). Hernandez et al. [51] found that the Sonoran propolis primarily consisted of phenolic acids, flavonoids, and their ester derivatives. Additionally, pinocembrin, chrysin, and pinobanksin-3-O-acetate were identified as the primary components in all three samples [51]. Specifically, rutin, naringenin, and hesperetin were exclusively present in propolis from PAP. Xanthomicrol was discovered in both PAP and CP samples, but the molecule 3'-desmethoxysudachitin was only detected in CP. Wollenweber and Buchmann [89] have identified the presence of xanthomicrol and 3′-desmethoxysudachitin as distinctive features of A. deltoidea exudates. This plant is widely distributed throughout the Sonoran Desert and may be associated with the botanical origin of CP and PAP. However, the only place where caffeic acid phenetyl ester (CAPE) was found was in UP. UP is a chemical compound that is typically found in propolis from temperate zones. Additionally, this sample contained higher levels of pinocembrin, chrysin, and pinobanksin-3-O-acetate, which are also commonly found in temperate propolis. This suggests that poplars from section Aigeiros, specifically P. fremontii, could be the source of this propolis [14, 55, 73].

Additionally, the chemical constitution of propolis from CP was further investigated in another study carried out by Li et al. [54], which resulted in the NMR characterization of three new flavonoids: (2*R*,3*R*)-3,5-dihydroxy-7-methoxyflavanone 3-(2-methyl)butyrate, (7″*R*)-8-[1-(4′-hydroxy-3′-methoxyphenyl)prop-2-en-1-yl] chrysin, and (7″*R*)-8-[1-(4′-hydroxy-3′-methoxyphenyl)prop-2-en-1-yl]galangin, and as well other 41 isolated chemical compounds characteristic of exudates from the genus *Populus*, including aromatic acids, flavonoids and its esters. In addition, the *in vitro* cytotoxicity of the 44 chemical compounds was evaluated on PANC-1 human pancreatic cell line, and (7′*R*)-8-[1-(4′-hydroxy-3′-methoxyphenyl)prop-2-en-1-yl]galangin showed to possess the most potent preferential cytotoxicity (PC_{50}: 4.6 μM) [54]. Lately, two phenylallylflavanones, (2*R*,3*R*)-6-[1-(4′-hydroxy-3′-methoxyphenyl)prop-2-en-1-yl]pinobanksin and (2*R*,3*R*)-6-[1-(4′-hydroxy-3′-methoxyphenyl)prop-2-en-1-yl]pinobanksin 3-acetate were identified in CP by first time and were new for propolis in general. These phenylallylflavanones additionally displayed a cytotoxic effect against PANC-1 (PC_{50}: 17.9 and 9.1 μM, respectively) [92]. Moreover, the cytotoxic evaluation of 39 of those 44 compounds, isolated from CP, was carried out on a panel of six different cancer cell lines: murine colon carcinoma (colon 26-L5), murine melanoma (B16-BL6), murine Lewis lung carcinoma (LLC), human lung adenocarcinoma (A549), human cervix adenocarcinoma (HeLa) and human HT-1080 fibrosarcoma (HT-1080). The compounds (2*R*,3*S*)-8-[4-Phenylprop-2-en-1-one]-4′,7-dihydroxy-3′,5-dimethoxyflavan-3-ol, cinnamyl *p*-coumarate and 2-acetyl-3-caffeoyl-1-*p*-coumaroylglycerol exhibited the most potent cytotoxic effect in comparison with the tested flavonoids, phenolic acid derivatives and glycerides from CP [53].

More studies have been done in order to determine the biological activities of Mexican propolis samples collected in semi-arid and arid zones (UP, PAP and CP). Sonoran propolis showed a strong antiproliferative effect on human and murine cancer cell lines A549 (IC50: 58.6 μg/mL), HeLa (IC50: 31.7–49.8 μg/mL), LS-180 (IC_{50}: 53.3–84.9 μg/mL), RAW 264.7 (Abelson murine leukemia virus transformed macrophages; IC_{50}: 0.8–5.2 μg/mL) and M12.C3.F6 (murine B-cell lymphoma cells; IC_{50}: 3.1–6.8 μg/mL). Moreover, CAPE, galangin, xanthomicrol and chrysin induced a significant antiproliferative effect on most of the cancer cell lines evaluated (IC_{50}: 3.2–95.4 μM) [51]. Since DNA harvested from cancer cells treated with UP exhibited a ladder of internucleosomal DNA cleavage pattern characteristic of apoptosis, in addition to the morphological changes observed in treated cells, a study conducted with the aim to determine biochemical events produced at earlier stages of apoptosis has been done. By annexin V-FITC/Propidium iodide double labelling, it has been demonstrated that Sonoran propolis treatment induced antiproliferative effect on M12.C3.F6 cells through apoptosis induction, and this apoptotic effect resulted to be mediated by modulations in the loss of mitochondrial membrane potential and through activation of caspases signalling pathway (3, 8 and 9). Additionally, some of the constituents of Sonoran propolis that induce apoptosis in cancer cells were characterized by an HPLC–PDA–ESI-MS/MS analysis, followed by isolation procedures and NMR spectroscopy that yield eighteen flavonoids, commonly described in poplar-type propolis, including two esters of pinobanksin, pinobanksin-5-methylether-3-*O*-propanoate and pinobanksin-5-methylether-3-*O*-butyrate were described by first time in propolis samples in general. Moreover, pinobanksin, pinobanksin-3-*O*-propanoate, pinobanksin-3-*O*-butyrate, pinobanksin-3-*O*-pentanoate, galangin, chrysin and CAPE induced antiproliferative activity on M12.C3.F6 cells through apoptosis induction [52].

The antibacterial and FRS activities of Sonoran propolis (UP, PAP and CP) have been tested by broth microdilution method and by DPPH assay, respectively. Sonoran propolis exhibited antibacterial activity against only Gram-positive bacteria, and UP presented the highest inhibition against *Staphylococcus aureus* (MIC: 100 μg/mL), followed by CP. CAPE, an exclusive constituent of UP, showed high growth inhibitory activity towards Gram-positive bacteria, particularly against *S. aureus* (MIC: 0.1 mM). CP presented the highest FRS activity (86% at μg/mL). The chemical constituents CAPE and rutin presented a high antioxidant activity (90.4 and 88.5% at 70 μM, respectively) in comparison with ascorbic acid control (95.0% at 70 μM). These results suggested that the presence of CAPE and rutin could be implied in the biological activities induced by Sonoran propolis [31]. Furthermore, the anti-*Vibrio* activity of those propolis samples collected in North-western Mexico was evaluated by broth microdilution method. UP presented the highest antibacterial effect against *Vibrio cholerae* O1 serotype Inaba, *V. cholerae* non-O1, *V. vulnificus* (MIC_{50}: <50 μg/mL), and *V. coholerae* O1 serotype Ogawa (MIC_{50}: 100 μg/mL). The constituents CAPE and galangin presented a potent growth inhibitory activity (MIC_{50}: 0.05–0.1 mM) against *V. cholerae* strains (non-O1 and serotype Ogawa) [93]. Additionally, the *in vitro* antiparasitic activity of Sonoran propolis against *Giardia lamblia* has been tested. UP showed the highest growth inhibitory effect (IC_{50}: 63.8 μg/mL) in comparison with CP and PAP (IC_{50}: >200 μg/mL). Among the chemical constituents of Sonoran propolis evaluated, CAPE had the highest growth inhibitory activity (IC_{50}: 222.1 μM), followed by naringenin (IC_{50}: 461.8 μM), hesperetin (IC_{50}: 494.9 μM) and pinocembrin activity (IC_{50}: 680.6 μM) [34].

Since UP showed to be one of the most biologically active of the Sonoran propolis tested, the evaluation of the seasonal effect on the chemical composition and biological activities (antiproliferative, antiparasitic and antioxidant activities) of UP has been done. The collected seasonal samples [spring (sp), summer (s), fall (f) and winter (w)] were analysed by an HPLC–DAD–UV method, wherein from the qualitative point of view, the chemical profile of the seasonal samples was similar; however, the results for antiproliferative effect on M12.C3.F6 cell line [sp (IC50: 11.6 μg/mL) > w (IC50: 26.6 μg/mL) > s (IC50: 49.7 μg/mL) > f (IC50: 54.5 μg/mL)] and antiparasitic activity on *G. lamblia* [s (IC_{50}: 23.8 μg/mL) > w (IC_{50}: 59.2 μg/mL) > sp (IC_{50}: 102.5 μg/mL) > f (IC_{50}: 125.0 μg/mL)] presented significant differences, which suggested that slightly quantitative variations on the bioactive constituents could be implicated in the seasonal effect of biological activities of UP. All propolis samples had weak FRS activity (<25% at 100 μg/mL) [34, 94].

Recently, the immunomodulatory properties of UP were tested (0.2–20.0 μg/mL) in a comparative study with Brazilian (Botucatu, Sao Paulo) and Cuban (Havana) propolis on pro- and anti-inflammatory cytokine production [tumor necrosis factor (TNF)-α and interleukin (IL)-10, respectively] by human monocytes. Brazilian propolis stimulated both TNF-α and IL-10 production by monocytes; meanwhile, Cuban propolis stimulated TNF-α and inhibited IL-10 production. UP exerted the opposite effect, inhibited TNF-α and stimulated IL-10 production. These results are due to qualitative and quantitative differences in the chemical constitution of the three samples, since different constituents that may exert pro- and anti-inflammatory activity depending on concentration. It is reported that the major compounds found in Brazilian, Cuban and Mexican propolis samples used in this study are artepillin C, isoflavonoids and pinocembrin, respectively, [46].

In a recent chemical comparative study performed with the aim to develop and validate a suitable RP–HPLC method to determine and quantify flavonoid markers in Mexican propolis, 11 samples collected at different ecological regions in six states (Estado de Mexico, Puebla, Chiapas, Zacatecas, Tlaxcala and Guanajuato) were analysed. Acacetin, 4′,7-dimethyl naringenin and 4′,7-dimethyl apigenin were used as marker components in this study, and the method was applied to establish some quantitative variations related to seasonal and geographical conditions of the propolis samples. 4′,7-Dimethyl apigenin was selected as an appropriate marker of

Mexican propolis, followed by 4′,7-dimethyl naringenin. Both chemical compounds were considered useful for quality control procedures in the geographical origin validation of Mexican propolis [95].

2.4. Cuba

Cuba is the largest island in the Caribbean Sea and is part of the Cuban Archipelago, which includes the Island of Youth and over 4000 other islands. Cuba's moderate tropical climate, combined with its exposure to various wind currents, topographical variations, diverse moisture levels, and different types of soil, results in the formation of heterogeneous ecological regions. These regions include wetlands along the southern coast, tropical desert-like conditions along the eastern coast, and pine forests in the mountains. As a result, Cuba showcases almost every ecosystem found in the Antilles Islands [97]. The Cuban propolis is the most extensively studied kind of propolis in North America, which includes the Caribbean and Central America. Numerous studies have been conducted to examine its composition and qualities.

The initial investigations conducted on Cuban propolis mostly concentrated on its biological qualities, as it is regarded as a customary homemade medicine in this nation. Nevertheless, the majority of these investigations were conducted without specific knowledge of the chemical composition of Cuban propolis. This lack of information makes it challenging to establish a clear link between the reported bioactivity from 20 years ago and the currently recognized chemical elements [58]. The antioxidant properties of ethanolic extracts of Cuban propolis from Baracoa and Pinar del Rio province were examined by assessing their ability to scavenge various oxygen radicals (superoxide and alkoxy) using luminol-sensitized chemiluminescence. Both preparations of propolis exhibited significant antioxidant activity against superoxide (IC50: 5.0 and 9.5 μg/mL, respectively) and alkoxy (IC50: 0.5 and 0.6 μg/mL, respectively) radicals [98].

Moreover, the hepatoprotective effects of Cuban red propolis (CRP) from Havana region were evaluated in different models of acute liver injury induced by paracetamol (600 mg/kg) [99, 100], and by allyl alcohol (64 mg/kg) [101], both in mice. The intraperitoneal administration of ethanolic extract of propolis (25, 50 and 100 mg/kg) showed to decrease significantly the enzymatic activity of alanine aminotransferase (ALAT), the levels of reduced gluthatione (GSH), and as well showed to reduce liver damage, these protective effects of propolis were produced both, before (30 min.) and after (2 h) paracetamol hepatotoxicity induction [99, 100]. Similar results were obtained with CRP administration before allyl alcohol (30 min.) [101].

Additionally, CRP was evaluated in other two models of acute hepatotoxicity induced by carbon tetrachloride (CCl_4) and by galactosamine (1000 mg/kg) in Sprague–Dawley rats [102–104]. The treatment with CRP (5, 10 and 25 mg/kg) in rats with hepatotoxicity induced by CCl_4, showed to reduce ALAT and hepatic malondialdehyde (MDA) levels in blood serum, and as well decreased the triglyceride (TG) levels in liver in comparison with control group [102]. Furthermore, histopathological evaluation revealed rats treated with CRP (25, 50 and 100 mg/kg) exhibited a significant reduction in liver injury, according to the low count of affected cells and the limited extension of steatosis area in comparison with those rats with control treatment [102, 103]. Similar biochemical histopathological results were obtained by CRP against the hepatotoxicity induced by galactosamine [104]. After these experiments, it was suggested that CRP probably exerted its hepatoprotective effects by antioxidant properties (scavenging action against oxygen radicals) [100–104].

The antipsoriatic, anti-inflammatory and analgesic effects of CRP were additionally assessed in another study, wherein CRP induced the formation of granular layer in a mouse tail model, reflecting its antipsoriatic activity; meanwhile, its anti-inflammatory effect was observed using three different test models [Cotton-pollot granuloma assay in rats Sprague–Dawley (dose: 50 mg/kg i.g.), croton oil-induced edema (dose: 25%; 2.5 μL) and the peritoneal capillary permeability test in Swiss albino mice (dose: 10 mg/kg)]. CRP exhibited an analgesic activity using the models of acid-acetic-induced writhing (25 mg/kg i.g.) and hot plate test in Swiss albino mice (40 mg/kg) [105]. Moreover, the sensitizing properties of CRP were assessed, and CRP did not induce erythema, edema, and the study also revealed the absence of dermal and ocular toxicity in guinea pigs and New Zealand rabbits at 24 h. However, a moderate contact allergy potential was identified in CRP treatment by slight induction of erythema [105].

Since Cuban propolis presented several biological activities, subsequent studies came as consequence of the renewed interest in exploring the chemical composition of propolis for drug development [61, 106]. An ethanolic extract of Cuban propolis collected from Nuevitas, Cuba has yielded the isolation and structural determination of propolone A, the first polyisoprenylated benzophenone isolated from tropical propolis [106]. Although the presence of polyisoprenylated benzophenones in propolis from the tropical areas has been previously proposed, the full characterization of individual benzophenone molecules has not happened until this study [107]. In addition, propolone A showed significant antimicrobial and fungicidal activities against several Actinomyces, yeasts and Gram-positive and Gram-negative bacteria [106] that encouraged the study and isolation of other bioactive prenylated benzophenone derivatives present in Cuban propolis.

The presence of polyisoprenylated benzophenones in Cuban propolis suggested that resins from genus *Clusia* could be implied as botanical source [108]. Copey tree (*Clusia rosea*) is widespread distributed throughout Cuba, and nemorosone, a prenylated benzophenone, is one of the major compound present in floral resins of copey tree [108, 109]. In that sense, a phytochemical study of 21 propolis samples collected from different locations along the country was carried out with the aim to track the presence of nemorosone and other benzophenones characteristic of *Clusia* species. Nemorosone was identified as major compound in propolis from the western, eastern and central Cuba. Additionally, a mixture of xanthochymol and guttiferone E, present in resins secreted by fruits but not on floral resins of *C. rosea*, was detected in a lesser proportion in those propolis samples. This study has also provided evidence about the cytotoxic effect of nemorosone on human cancer cell lines HeLa, Hep-2, PC-3 and U251 (IC_{50}:1.9–7.2 μM) and FRS activity (IC_{50}: 44.1 μM). Nemorosone showed to be more biologically active than a mixture of two its methyl derivatives on cytotoxicity (IC_{50}: 29.4–94.5 μM) and antioxidant (IC_{50}: >200) evaluations [109]. Later, comprehensive chemical composition analysis of propolis collected from Guantánamo region led to the structural determination of three new polyprenylated benzophenones derivatives (propolones B-D), in addition to the presence of garcinielliptone I and hyperibone B [110].

By virtue of the differences of endemic plants in tropical and temperate regions and also the ecological diversity of Cuba, further investigations on the chemical composition of propolis from different regions of Cuba were carried out. A scrupulous chemical analysis of a CRP from Pinar del río provided a cutting edge input on composition of propolis in general by reporting for the first time the presence and structural determination of isoflavonoids (isoflavones, isoflavans and pterocarpans) in a propolis sample, in addition to gallic acid, isoliquiritigenin, and (-)-liquiritigenin [57]. The occurrence of isoflavonoids is pretty restricted in nature and is characteristic of Leguminosae family, and these findings yield two different Cuban propolis types, suggesting the participation of at least two diverse botanical sources, one that provides prenylated benzophenones and another the isoflavonoids [57].

With the aim to study the chemical similarities and differences on Cuban propolis and to establish a classification system according to the presence of secondary metabolites, 65 propolis samples from different regions of Cuba were chemically analysed by HPLC–PDA, HPLC–ESI/MS and NMR. Cuban propolis types were grouped in three groups, including CRP rich in isoflavonoids; Cuban brown propolis (CBP), characterized by a high amount of polyisoprenylated benzophenones; and Cuban yellow propolis (CYP) that contains aliphatic compounds, which was sub-classified depending on its content on triterpenic alcohols (type A) and polymethoxylated flavonoids (type B) [58, 111–113]. Once CBP was identified as the major type of propolis produced in Cuba, biological assays were performed, especially on its capacity to inhibit *in vitro* cancer cell proliferation, since nemorosone, its main constituent, showed a potent cytotoxic effect on human cancer cells [109, 111]. CBP exhibited anti-metastatic effect in mouse mammary carcinoma Ehrlich's ascites tumour (EAT) cells in NMRI immunocompetent mice (5–23 μg/mL), in addition to cytotoxicity on diverse cancer cell lines, suggesting the potential of Cuban propolis as a source of possible anticancer agents [114].

Other studies showed that BCP induced significant antiproliferative activity on human breast cancer cell lines, preferentially on MCF-7 (estrogen receptor positive; ER α+) rather than MDA-MB-231 (estrogen receptor negative ER α-) in a dose-dependent manner. The antiproliferative effect on MCF-7 was partially related to apoptosis induction after an arrest in G1 phase of cell cycle was detected. Moreover, the co-administration of 17-β-estradiol and an antagonist (ICI 182,780) allowed to hypothesize that BCP possesses an estrogen-like activity, although the effect would not be exclusively considered ER-dependent because the mortality also induced on MDA-MB-231 by BCP [115].

Otherwise, the treatment with nemorosone on MCF-7, MDA-MB-231 and LNCaP cells induced a selective antiproliferative effect on MCF-7 by arresting the cell cycle in G0/G1 phase, in addition to a reduction in the expression of pERK1/2 and PAkt. Nemorosone did not induce antiproliferative effect on MDA-MB-231 nor human prostate cancer cells LNCaP (which express ERβ but not ERα), which suggested that nemorosone is the main responsible for the antiproliferative effect of BCP on ERα+ breast cancer cells, and it could have therapeutic applications in breast cancer treatment since its activity on ERα+ cells [116]. Moreover, nemorosone exhibited cytotoxicity in neuroblastoma cell lines, including their clone selected for resistance to chemotherapeutic compounds. It induced a G0/G1 arrest on cell cycle that yields a reduction in S-phase population, in addition to the detection of an upregulation of $p21^{Cip1}$, presence of apoptotic DNA laddering, the activation of caspase 3 activity, dephosphorylation of ERK1/2 in LAN-1 and the inhibition of Akt/PKB [114]. Due to the reported correlation of nemorosone cytotoxicity on cancer cell lines to direct action on estrogen receptor (ERs), other studies have been done. By *in vitro* tests [recombinant yeast assay (RYA) and E-screen assay], the antiestrogenic activity of nemorosone was demonstrated, exhibiting this benzophenone a reduction on the cell proliferation induced by 17-β-estradiol (E2). Additionally, the treatment with nemorosone did not induce DNA damage in breast cancer cells MCF-7 BUS or in normal breast cells MCF10A. These results suggested that nemorosone could be a promising adjuvant for ER antagonists [117].

In addition to the presence of nemorosone in CBP, there have been identified mucronulatol and plukenetione A as cytotoxic and antiproliferative compounds that could contribute to the potent inhibitory activity of CBP on cancer cell proliferation. The occurrence of plukenetione A has previously been identified in *Clusia plukenetii*, and in later studies in CBP. Plukenetione A is reported to exert considerably cytotoxicity in a panel of cancer cell lines, including colon, ovarian, prostate carcinomas and neuroblastoma cells (IC_{50}: between 1.7 and 16.3 µg/mL); in addition, it induced in HCT8 cells that the depletion of S phase transitory cells as consequence of a G0/G1 arrest in cell cycle, followed by the presence of apoptotic DNA laddering, changes in gene expression patterns of genes required for cell replication and maintenance, accompanied by the inhibition of the enzymatic activity of both topoisomerase I and DNA polymerase [118]. The isoflavonoid mucronulatol has been described as one of the most cytotoxic constituents for Caribbean propolis. In general mucronulatol showed cytotoxicity MDR1−/MDR3+ cells (2.7–10.2 µg/mL), but not on MDR1+ cells (> to 100 µg/mL), which resulted as consequence of an interruption of cell cycle progression, by blocking at G1, accompanied by an upregulation of $p21^{Cip1}$ and $p27^{Kip1}$ and a downregulation of cyclin E and CDK4, interfering in general with the cell cycle machinery [119]. The presence of mucronatol has also been reported in Brazilian and Mexican red propolis [56].

Other different biological properties of nemorosone have been evaluated, including the mutagenic, antimutagenic and estrogenic effects. The mutagenic and antimutagenic activity of nemorosone were assessed by the Ames test on *Salmonella typhimurium* strains (TA97a, TA98, TA100 and TA102), wherein nemorosone did not induce any mutagenic activity; meanwhile, nemorosome exhibited a moderate to strong protective effect (31 and 53% of inhibition, respectively) in association with mutagens in strains TA100 and TA102. Nemorosone induced estrogenic activity detectable by recombinant yeast assay at various concentrations (EEq of 0.41 ± 0.16 nM), concluding with those results that nemorosone could have a chemotherapeutic application in breast cancer research [120].

Red propolis has also been documented in additional tropical nations, including as Brazil, Mexico, and Venezuela. Brazilian red propolis (BRP) is a type of propolis that is made by honeybees using the resinous exudates of Dalbergia ecastophyllum in Northeastern Brazil [121, 122]. Its chemical composition is primarily composed of isoflavonoids, neoflavonoids, flavonoids, and polyisoprenylated benzophenones (PPBs) [123, 124], which indicates certain similarities to Cuban red propolis (CRP). To examine the chemical composition and botanical origin of various red propolis from the Americas, a comparative study was conducted using HPLC-DAD-MS analysis. Additionally, exudates from D. ecastophyllum (DEE) were included in the analysis [125]. Flavanones, isoflavones, isoflavans, pterocarpans, and a chalconoid were found in both red-type propolis (BRP and CRP) and DEE. Furthermore, only BRP included guttiferone E/xanthochymol and oblongifolin A. Both flavans, retusapurpurin A and the newly discovered retusapurpurin B, were identified as the pigments accountable for the red hue observed in those samples. The results suggest that propolis collected from several tropical regions in the Americas have comparable phytochemical compositions, indicating a common source of exudates from Dalbergia species, most likely DEE. Nevertheless, the existence of PPBs in BRP indicated that this sample likely has a supplementary plant-based source. This investigation yielded useful insights for the purpose of achieving a more accurate categorization of propolis [125].

All of these investigations contribute to the comprehension of the chemical variability found in propolis from tropical regions, where Populus species are absent and honeybees must collect bioactive secretions from alternative plant sources. Furthermore, these research offer a detailed analysis of the pharmacological properties of prenylated benzophenones and isoflavonoids. These chemical components show great potential as therapeutic agents and could serve as important indicators in the future standardization of Cuban propolis. Further investigations are required to comprehend the mechanism by which CRP exerts its hepatoprotective effects, as well as to identify the specific chemical components involved.

2.5. El Salvador

It has been noted that honeybees in the tropics had to search for new plant sources of bioactive resins compared to those they typically visit in temperate zones. This has led to a high level of chemical diversity in tropical propolis, which has sparked significant attention due to the pharmacological qualities of its constituents. Tropical propolis contains a range of compounds, such as polyprenylated benzophenones, isoflavonoids, and triterpenes in Cuban propolis [57, 111, 113]. Brazilian propolis also contains isoflavonoids and polyisoprenylated benzophenones. Mexican propolis, on the other hand, contains 1,3-diarylpropane derivatives, flavonoids, and isoflavonoids [56].

With the aim to continue the studies of propolis from tropical Central America, a sample collected in the vicinities of Usulutan, El Salvador, was studied and two new chalcones were isolated and characterized by NMR (2′,3′-dihydroxy-4,4′-dimethoxychalcone and 2′,3′ 4-trihydroxy-4′-methoxychalcone) by first time in propolis. Both compounds presented a good antibacterial activity on *S. aureus* (29 ± 3 and 23 ± 1 mm of inhibition zone, respectively) but not on *E. coli*. In addition, those chalcones inhibited *C. albicans* growth (19.3 ± 0.6 and 29 ± 1 mm, respectively); thus, better antibacterial and antifungal activities than original propolis extract were presented (12 ± 1 and 11 ± 1 mm, respectively). However, none of them exhibited a more effective cytotoxicity in brine shrimp (*A. salina* nauplii) lethality bioassay than propolis extract (LC_{50}: 39 ± 9 mg/mL) [126].

In another study, two diterpene glycosides (*ent*-8(17)-labden-15-*O*-α-$_L$-rhamnopyranoside and *ent*-8(17)-labden-15-*O*-(3′-*O*-acetyl)-α-$_L$-rhamnopyranoside) were isolated from a propolis sample from the Eastern region of El Salvador, and both were reported by the first time in propolis. These new labdenol glycosides exhibited good antibacterial activity against *S. aureus* (21.0 ± 1 and 20.3 ± 0.6 mm at 4 mg/mL, respectively) and showed to be more effective than the propolis ethanolic extract (12 ± 1 mm at 4 mg/mL), but those glycosides did not induce any effect on *E. coli* and *C. albicans*. Additionally, *ent*-8(17)-labden-15-*O*-(3′-*O*-acetyl)-α-$_L$-rhamnopyranoside exhibited a better cytotoxicity (LC_{50}: 15 ± 7 mg/mL) than the extract (LC_{50}: 39 ± 9 mg/mL) in brine shrimp lethality bioassay [127].

2.6. Honduras

Central America exhibits a significant abundance of unique species. Salvadoran propolis has been found to contain chalcones and diterpene glycosides, which are not present in propolis from other tropical regions such as Cuba, Brazil, and Mexico. To further understand the variations and similarities in tropical propolis across the Americas, additional samples from Central America were chemically analyzed. As far as we know, there has only been one study conducted on Honduran propolis. In this study, a sample collected in Marcala was separated into different parts, which resulted in the discovery of cinnamic ester derivatives, including a new (E,Z)-cinnamyl cinnamate. Additionally, flavanones, triterpenes, aromatic acids, and one chalcone were also found in the sample. Honduran propolis inhibited the ATPase activity of Pdr5p (70% at 100 µg/mL), a protein responsible for a multidrug resistance phenotype in *Saccharomyces cerevisiae*, and four of its most abundant constituents (*E*)-Cinnamyl-(*E*)-cinnamate (IC_{50}: 2.58 µM), (*E*)-Cinnamyl-(*E*)-*p*-coumarate (IC_{50}: 1.54 µM), 6β-Hydroxy-3-oxo-lup-20(29)-en-28-oic acid (IC_{50}: 1.03 µM) and sakuranetin (IC_{50}: 1.20 µM) were also potent inhibitors [88].

The presence of these cinnamic ester derivatives has been described in exudates and volatile fractions of *L. styraciflua* (Honduras styrax, Hamamelidaceae). Interestingly, high amount of the triterpene compound 6β-hydroxy-3-oxolup-20(29)-en-28-oic acid have been reported in the cones of *L. styraciflua*, which suggested that *L. styraciflua* would be the botanical source of this Honduran propolis sample, in addition to the relative abundance of this plant in the surrounding areas to the beehives [88]. Subsequent findings in North Carolina propolis revealed the existence of cinnamic ester derivatives and 6β-hydroxy-3-oxolup-20(29)-en-28-oic acid, along with the presence of distinctive poplar flavonoids. Thus, the exudates of L. styraciflua were suggested as an alternative botanical source for North Carolina propolis. Honduran and North Carolina propolis have a common plant origin, which challenges the prevailing idea that poplar-type propolis is only found in temperate areas like North Carolina. Instead, these propolis types also contain resins collected from plants in both tropical and temperate regions of North America [87, 88].

3. SOUTH AMERICAN PROPOLIS

South America is known for having the greatest plant diversity compared to any other region in the world. This is attributed to its large size, location (latitude and longitude), the presence of extensive tropical forests, and the Andes Mountains, which are the largest mountain system globally. The Andes Mountains serve as a connection between tropical and temperate latitudes across South America [128, 129]. An analysis has been conducted on propolis samples collected from several places in South America. The analysis of Brazilian propolis surpasses that of any other propolis in South America, with Argentinean, Chilean, Uruguayan, and Venezuelan propolis following suit. Currently, there has been research conducted on propolis from Colombia and Bolivia, but propolis from Peru has only been examined in biological comparison studies. This section provides a concise summary of the latest and most significant advancements in the field of South American propolis research.

3.1. Colombia

Colombian vegetation possesses a great biodiversity with a high number of endemic plant species distributed in a variety of tropical forests, steppe and grasslands, representing a source of great variety in phytochemical substances [130]. Propolis is used in Colombia as a folk remedy, and in cosmetic and food industries, however, few studies in Colombian propolis have been done [131]. In a recent study, propolis collected from Medellin region, and subsequently extracted with n-hexane/methanol and fractioned with dichloromethane, exhibited an inhibitory effect on mycelial growth against *Botryodiplodia theobromae* (23.5% at 1 mg/mL) and two different strains of *Colletotrichum gloeosporioides* (38.1 and 47.6% at 1 mg/mL, respectively), which are important postharvest fungi that affect tropical and subtropical fruits. Through antifungal bioassay-guided fractionation of dichlorometane fraction, three labdan-type diterpenes were characterized: isocupressic acid [15-hydroxylabda-8(17),13E-dien-19-oic acid], (+)-agathadiol [labda-8(17),13-diene-15,19-diol] and epi-13-torulosol [8(17),14-1abdadien-13*S*, 19-diol], chemical compounds that could be responsible of the antifungal effect of Colombian propolis [131].

The presence of isocupressic acid and other labdan diterpenes has been described in Brazilian green propolis, a propolis type that has its plant sources in the resins of *Baccharis* spp. and *Araucaria heterophylla* [23]. In addition, the occurrence of agathadiol, isocupressic acid and torulosol has been reported in Algerian propolis, where its main botanical origin seems to be *Cistus* spp. (Cistaceae) [132]. In temperate regions of America, including Central America, inhabit three genera of Cistaceae (*Crocanthemum* spp., *Hudsonia* spp. and *Lechea* spp.), however, in order to draw a phytochemical origin of those compounds, the botanical origin has to be investigated. Moreover, in another study, a propolis sample collected from La Union (Antioquia), showed a weak to moderate antifungal activity (MIC_{50}: >1 mg/mL) on *C. acutatum*, *C. gloeosporioides*, *Aspegillus* sp. and *Penicillium* sp., in addition to a bacteriostatic (1.0 mg/mL) and bactericidal (10.0 mg/mL) effect against *B. subtilis* (Gram+). By GC–MS were detected fatty acids, and their esters, sesquiterpenes, pentacyclic tripterpenes and bicyclic labdan-type diterpenes, such as isocupressic acid, agathadiol and 13-epi-turolosol [133]. The presence of terpenes in both analysed samples of Colombian propolis suggested a possible common botanical source that has to be investigated in order to understand the origin of those phytochemicals and the biological activities of Colombian propolis.

3.2. Venezuela

Like Colombia, Venezuela is a tropical region in South America that is home to a vast array of unique plants and a wide diversity of bioactive chemicals. A preliminary chemical compositional analysis of tropical propolis was conducted using 38 samples collected from various tropical regions in Venezuela, such as Bolívar, Nueva Esparta, Cojedes, Yaracuy, Monagas, Mérida, Táchira, Portuguesa, Sucre, Baringas, Aragua, and Carabobo. The objective was to identify phytochemical markers that could indicate the botanical source of Venezuelan propolis. The majority of these propolis samples had comparable chromatographic patterns that are indicative of phenolic chemicals, particularly prenylated benzophenones. Flavonoids, on the other hand, were only detected in a small number of samples, and they were recognized as methylation 6-oxygenated flavones. The High Performance Liquid Chromatography (HPLC) analysis of the resins released by the flowers of Clusia minor indicates a highly comparable phenolic profile. This suggests that the botanical origin of the polyprenylated benzophenones found in Venezuelan propolis is the exudates from C. minor and other Clusia species. This information is supported by reference [107].

Moreover, in a later study on Venezuelan propolis from Trujillo State, the presence of the already reported polyisoprenylated benzophenones scrobiculatones A and B was identified, together with their derivatives 18-ethyloxy-17-hydroxy-17,18-dihydroscrobiculatone A and 18-ethyloxy-17-hydroxy-17,18-dihydroscrobiculatone B (first time described in this study). Additionally, a mixture of scrobioculatones A-B exhibited a significant antibacterial activity against *S. aureus* (MIC: 125 µg/mL), a moderate toxicity towards *A. salina* nauplii (LC_{50}: 14 ± 6 µg/mL) and low FRS activity by DPPH assay (10%) [134]. The chemical composition and the antioxidant effect of other three samples (V-1, V-2 and V-3) collected in San Antonio de los Altos, Venezuela, was subsequently analysed. The presence of prenylated benzophenone derivatives, diterpenes and triterpenes was detected by GC–MS, suggesting again the participation of *Clusia* spp. resins in Venezuelan propolis collected from different regions. Additionally, propolis samples (0.04% wt.) increased the oxidation stability of triacylglycerol molecules in comparison with control, and the effect of V-2 and V-3 samples was higher than V-1 (approximately 1.5-fold) [135].

3.3. Peru

Chemical analysis of Peruvian propolis has not been conducted yet. However, a comparative study was performed to evaluate the cytotoxic, hepatoprotective, and FRS (DPPH method) properties of propolis samples collected from Peru, Brazil, Netherlands, and China. For each propolis sample, extracts of methanol (MeOH ext.) and water (Wt ext.) were produced.

Peruvian propolis showed the lowest FRS activity among the other propolis samples (ED_{50}: MeOH ext. 82.3 μg/mL and Wt ext. 94.9 μg/mL), although it exhibited moderate cytotoxicity on murine colon 26-L5 carcinoma (51.1 μg/mL) and HT-1080 fibrosarcoma cells (76.4 μg/mL). The hepatoprotective activity of propolis was tested on primary cultures of mouse hepatocytes induced to cell death with D-galactosamine (D-GalN) and tumor necrosis factor-α (TNF-α), wherein the MeOH extract of Peruvian sample showed toxicity at 200 μg/mL; meanwhile, the water extract exhibited significant hepatoprotective activity at 200 μg/mL (approximately 45%) [136].

3.4. Bolivia

Bolivia is situated in the heart of South America and its land encompasses a transition between wet tropical and dry subtropical climate regions. The country's geographical qualities contribute to a significant presence of biodiversity. However, there is a lack of phytochemical studies conducted on the flora of Bolivia. In Bolivia, propolis is utilized as an antibacterial medication for the treatment of infections and respiratory ailments [137]. Recent research has focused on Bolivian propolis, specifically analyzing the chemical composition of ten samples obtained from the primary beekeeping locations in Bolivia. This investigation has revealed the existence of two distinct forms of propolis. The first group refers to propolis obtained from the valley regions (Cochabamba, Chuquisaca, and Tarija). This propolis is characterized by a significant presence of prenylated phenylpropanoids, including p-coumaric and caffeic acid derivatives such as drupanin (3-prenyl-p-coumaric acid) and artepillin C (3,5-diprenyl-p-coumaric acid). These compounds have been previously identified in Brazilian green propolis derived from Baccharis dracunculifolia exudates. [137, 138]. Meanwhile, the second type of Bolivian propolis came from La Paz and Santa Cruz regions, and their chemical composition was mainly comprised by cycloartane and pentacyclic triterpenes, including cycloart-24-en-3β,26-diol, cycloart-24-en-3-one, cycloart-24-en-26-ol-3-one, mangiferonic acid methyl ester and lup-20(29)-en-3-one, which were triterpenes identified for first time in propolis in general. The antioxidant capacity of these ten samples was determined by DPPH, FRAP and ABTS assays, propolis samples rich in phenolic compounds presented moderate to strong antioxidant activity, while propolis rich in triterpenes showed a weak active [137].

Moreover, the antimicrobial properties of those ten Bolivian propolis samples and their main constituents were tested against 11 bacterial pathogenic strains of clinical relevance (*S. aureus* ATCC 25923, methicillin-resistant *S. aureus* ATCC 43300, *Escherichia coli* ATCC 25922, the clinical isolated *E. coli* 121, *E. coli* 122, *E. coli* LM2, *Salmonella enteritidis* MI, *Salmonella* sp. (LM), *Yersinia enterocolitica*-PI, *Pseudomonas* sp. and *Proteus mirabilis* 94-2) using micro-broth dilution method. Propolis samples exhibited different effects, from inactive (MICs > 1000 μg/mL) to low (MICs 250–1000 μg/mL), moderate (62.5–125 μg/mL) and high antibacterial activity (MIC 31.2 μg/mL). The samples that were rich in phenolics showed the high antibacterial activity; in comparison, terpene-rich samples were mostly inactive and some presented low activity. Kaempferol-3-methyl ether and drupanin were the most active constituents of Bolivian propolis. Additionally, the activity of propolis samples towards promastigotes of *Leishmania amazonensis* and *L. braziliensis* was evaluated and the results obtained were similar to those of antibacterial assays. The most active samples against *L. amazonensis* and *L. brasiliensis* were from Cochabamba (IC_{50}: 12.1 and 7.8 μg/mL, respectively) and Tarija (IC_{50}: 8.0 and 10.9 μg/mL, respectively) [139]. These studies offered a glimpse into the variations of phytochemicals in propolis from various regions dedicated to beekeeping in Bolivia. The findings suggest the presence of at least two distinct botanical resins, which need to be characterized in order to link the pharmacological activities of the propolis samples to specific plant origins. The results indicated that Bolivian propolis, which is rich in phenolic compounds, has the potential to be a valuable source of medicinal agents with antibacterial and antiparasitic properties.

3.5. Brazil

Brazil possesses a vast terrain that encompasses the most abundant and diverse plant life on Earth, with over 56,000 species, accounting for around 19% of the world's total flora [140]. Brazilian propolis exhibits a unique chemical diversity, encompassing many plant resins from distinct ecological regions. This leads to the formation of diverse varieties of propolis, which include typical chemical compositions seen in both tropical and temperate locations. Currently, Brazilian propolis is the subject of extensive research and numerous chemical components have been found in samples taken from various regions of the country. In this article, we want to provide a concise overview of the extensive chemical and pharmacological research conducted on Brazilian propolis by presenting the most noteworthy findings from numerous investigations.

One of the first studies, focused on chemical characterization of bioactive constituents of Brazilian propolis, was carried out with a sample collected in Sao Paulo state, which yielded the isolation of three new compounds: 3,5-diprenyl-4-hydroxycinnamic acid, 3-prenyl-4-dihydrocinnamoloxycinnamic acid and

2,2-dimethyl -6-carboxyethenyl-2H-1-benzopyran [141]. These Brazilian propolis constituents presented antimicrobial activity against *Bacillus cereus* (MIC 15.6, 31.3 and 125 μg/mL, respectively), *Enterobacter aerogenes* (31.3, 62.5 and 125 μg/mL, respectively) and towards *Arthroderma benhamiae* (15.6, >250 and 62.5 μg/mL, respectively). Although the occurrence of cinnamic acid and other phenylpropanoid acid derivatives has been previously reported in propolis from temperate regions, the finding of prenylated derivatives was new for propolis [141].

In other studies, the occurrence of fifteen more cinnamic acid derivatives, in addition to nine *p*-coumaric acid derivatives, was identified from two different samples collected in Minas Gerais, together with other compounds, including six caffeoyl quinic acid derivatives, nine flavonoids, one prenylated phenolic acid, four diterpenoic acids and one lignin [142, 143]. Additionally, benzofuran derivatives (A and B) together with two known isoprenylated compounds were isolated from Brazilian propolis (not specified region). A moderate cytotoxicity was showed by benzofurans A and B against highly liver-metastatic murine colon 26-L5 carcinoma (12.4 and 13.7 μg/mL) and human HT-1080 fibrosarcoma cells (13.9 and 43.2 μg/mL) [144]. At present, hundreds of studies have been performed in order to chemically characterize the different samples of Brazilian propolis, revealing the great variety of compounds in those samples, including more than 145 constituents that have been identified and some of them were additionally tested to characterize their pharmacological activities [28, 58].

In order to establish the basis of Brazilian propolis classification, 500 samples were collected from different regions, including southern (Paraná and Rio Grande do Sul), Southeastern (Sao Paulo, Minas Gerais) and Northeastern (Bahía, Piauí, Ceará and Pernambuco) of Brazil. According to their chemical profile, the samples were classified into 12 groups, wherein five groups were identified in the Southern, one from Southeastern and six from Northeastern Brazil, suggesting a greater plant diversity on Northeastern and Southern than Southeastern region [145, 146]. The antimicrobial activity against *S. aureus* and *S. mutans* was tested, and propolis from group 6 originated from Bahia state (Northeastern Brazil) showed the highest inhibitory effect on both microorganisms (inhibition zone: 6 and 9 mm, respectively) [146]. Subsequently, the botanical origin of one representative group of each region was investigated. The foraging behaviour of honeybees that produced Brazilian brown propolis (BBP) from the South (Paraná, group 3), Brazilian red propolis (BRP) collected in Northeastern (Bahia, group 6) and Brazilian green propolis (BGP) collected from Southeastern (Sao Paulo, group 12) was observed to identify which plant is visited to gather the respective resins. The plant buds or unexpanded leaves were cut off to extract the resinous material, and using reversed-phase high-performance thin-layer chromatography (RPHPTLC), reversed-phase high performance liquid chromatography (RPHPLC), and GC–MS, the plant origin was revealed [145].

The main compounds identified in propolis from group 3 were flavonoids and organic acids, and the chemical profile of that sample was similar to that of poplar exudates, in particular to *Populus alba*, which were assumed as the main botanical source of this propolis from Paraná state. Interestingly, poplar trees are not native in Southern South America; nevertheless, European immigrants planted poplar trees in the temperate Southern Brazil [71, 147]. In propolis from group 6, the presence of some aromatic compounds, terpenoids and fatty acid esters was recognized and exudates from *Hyptis divaricata* Pohl were suggested to be the main botanical source according to the chromatographic profile [71]. However, since some disparities were observed in the chemical profile of *H. divaricata* and propolis of group 6, in addition to the identification of prenylated benzophenones (hyperibone A) in a later study, both facts suggested the participation of resins from some species of *Clusia* genus as supplementary botanical source [58, 148].

In group 12, the presence of prenylated derivatives of *p*-coumaric acid, the characteristic artepillin C (3,5-diprenyl-4-hydroxycinnamic acid), among other aromatic acids (dihydrocinnamic, *p*-coumaric, ferulic, caffeic and caffeoylquinic acids) and some flavonoids, including kaempferid, 5,6,7-trihydroxy-3,4′-dimethoxyflavone and aromadendrine-4′-methyl ether, and other compounds were found. Based on chemical evidence, *Baccharis dracunculifolia* resins resulted to be the main botanical source of this propolis type, in addition to *Araucaria angustifolia* and *Eucalyptus citriodora* exudates [59, 71]. The antibacterial and antifungal effect of propolis from group 12 against *S. aureus* and *C. albicans* (10.5 and 15 mm of inhibition zone, respectively) allowed to determine that the most bioactive resins came from *B. dracunculifolia* leaf exudates (9 and 16 mm, respectively) [59]. Subsequent studies on BGP and *B. dracunculifolia* exudates (Minas Gerais, group 12) by HPLC–APCI–MS and GC–MS allowed to identify 126 constituents present in those samples, including mainly cinnamic acid and its derivatives, flavonoids, benzoic acid and a few benzoates, non-hydroxylated aromatics, and aliphatic acids and esters, in addition to prenylated compounds, alkanes and terpenoids [138]. Moreover, particular triterpenoids have been identified in BGP, including lupeol, esters of lupeol, α- and β-amyrins, cycloartenol, lanost-7,24-dien-β-ol [58].

With the aim to establish a more precise correlation among Brazilian propolis types, thirty-eight samples were collected, from Minas Gerais, Sao Paulo, Paraná, Mato Grosso do Sul, Bahia and Algoas, and were analysed by ESI-MS and tandem mass spectrometry (ESI-MS/MS). A principal component analysis (PCA) has been applied to determine similarities and differences in the secondary metabolite fingerprint of propolis samples. Brazilian propolis was divided into groups, according to their chemical profile and geographical origins. BRP from Algoas and Bahia was divided into two main groups (R1 and R2), and as well propolis from South and Southeastern Brazil were divided in one BGP group, which contained the greatest number of samples, and two groups of BBP (B1 and B2) from Paraná state. Seven compounds were used as markers and allowed the classification of Brazilian propolis into those five groups according to PCA analysis, wherein the presence of intersection of constituents among samples was detected. Pinocembrin was in BRP R2 and BBP B1, and chsysin in BBP B1, 2,2-dimethyl-6-carboxyethenyl-2H-1-benzopyran in BBP B2, 3-prenyl-4-hydroxycinnamic acid, 3,5-diprenyl-4-hydroxycinnamic acid and dicaffeoylquinic acid were identified in BGP and BBP B2, *p*-coumaric acid in BGP and BBP (B1 and B2), and 3-methoxy-4-hydroxycinnamaldehyde in BBP (B1 and B2). As observed, there are chemical crossing in the five groups [84].

In regard to the chemical composition of BRP, recent studies have reported the presence of prenylated benzophenones, such as guttiferone E/xanthochymol mixture, the isoflavonoids isosativan and medicarpin, the triterpenoid ketone 20(29)-lupen-3-one, and also a naphthoquinone epoxide (isolated for the first time from a natural source) in a sample collected from the state of Algoas [124]. The presence of these compounds suggested a mixed plant origin, since prenylated benzophenones are described in floral resins of *Clusia* species (Clusiaceae); in particular, the mixture of guttiferone E/xanthochymol and the triterpenoid have been reported in Cuban propolis [58]. Meanwhile, isoflavonoids have been related to *Dalbergia* species (Leguminosae). Additionally, isosativan, medicarpin and the guttiferone E/xanthochymol mixture presented antibacterial activity by inhibiting *S. aureus* (14, 23 and 19 mm), *E. coli* (0, 14 and 12 mm) and *C. albicans* (15, 26 and 0) growth. Moreover, the mixture guttiferone E/xanthochymol exhibited FRS activity by DPPH assay (49% at 48 μM) [124]. It is also reported that the presence of retusapurpurins A and B (C30 isoflavanes) provides the red pigment to those red propolis in different regions of Americas [125].

More studies on chemical composition of BRP collected from Algoas have been done, identifying methyl o-orsellinate, methyl abietate, medicarpin, 2,4,6-trimethylphenol, and the isoflavonoids, homopterocarpin, 4′,7-dimethoxy-2′-isoflavonol and 7,4′-dihydroxyisoflavone. Biological tests have been carried out for the BRP extract, including antimicrobial against *S. aureus* ATCC 25923 and *S. mutans* UA159 (MIC: 25–50 μg/mL), antioxidant (57% at 90 μg/mL) and cytotoxic activity against HeLa cell line (IC_{50}: 7.45 μg/mL) [149]. In another study of BRP, the cytotoxicity of a sample collected from the South coast of Paraiba State was evaluated on Human pancreatic cancer cells (PANC-1), and since Paraiba BRP exhibited a 100% cytotoxicity at 10 μg/mL, a subsequent phytochemical analysis was carried out and led to the isolation of 43 compounds, mainly flavonoids, including pterocarpanes, flavanonols, isoflavanones, chalcones, isoflavans, isoflavones, flavanones, lignans, a flavonol, a isoflavanonol and a neoflavonoid. Three novel compounds: (6aS,11aS)-6a-ethoxymedicarpan, 2-(2′,4′-dihydroxyphenyl)-3-methyl-6-methoxybenzofuran and 2,6-dihydroxy-2- [(4-hydroxyphenyl)methyl]-3-benzofuranone, in addition to

(6aR,11aR)-3,8-dihydroxy-9-methoxy-9-methoxypterocarpan, exhibited the most potent effect (100% cytotoxicity at 12.5 μM) on PANC-1 cells [123].

The chemical composition of an essential oil extract obtained from Brazilian propolis collected in Rio de Janeiro was characterized, and 26 constituents were identified, wherein β-caryophyllene (12.7%), acetophenone (12.3%) and β-farnasene (9.2%) were the major constituents, in addition to the new compounds found linalool (6.47%), Υ-elemene (6.25%), methyl hydrocinnamate, ethyl hydrocinnamate, α-ylangene and valencene. This essential oil extract presented antimicrobial activity against *S. aureus* (14 mm, inhibition zone), *S. epidermidis* 25/04 (10 mm), *S. epidermidis* 194/02 (10 mm), *S. pyogenes* 93007 (14 mm), *S. pyogenes* 75194 (18 mm) and *E. coli* (17 mm) by agar diffusion method [150].

Since Brazilian propolis is continuously produced all over the year, it was important to determine the seasonal effect on propolis composition and evaluate the influence of season on its biological activities. By GC and GC–MS, the seasonal variations in chemical composition of BGP collected from Sao Paulo state were investigated. All the seasonal samples contained phenolic compounds, mainly cinnamic acid derivatives as major constituents; however, the presence of diterpenes appeared in summer sample and reached a predominate percentage in the autumn sample, but being absent during the other seasons, which suggested the participation of at least two plants as botanical source [59]. These chemical variations could be important for the practical application of propolis, and in order to evaluate the effect of these slight changes on Brazilian propolis composition, antibacterial assays on bacterial strains isolated from human infections were carried out. The growth of Gram-positive bacteria, *S. aureus*, was inhibited at low concentrations of seasonal propolis samples (0.4% v/v), whereas Gram-negative, *E. coli*, bacteria were less susceptible to propolis treatment. It was concluded that there was no significant difference on the inhibitory action of propolis samples, discarding a seasonal effect on antibacterial activity of propolis from Sao Paulo, Brazil [30].

Propolis possesses immunomodulatory activities, and it has been reported that its immunostimulant effect is produced via macrophage activation, which enhances macrophage phagocytic capacity [28, 151, 152]. The effect of BGP on activation by reactive oxygen (H_2O_2) and nitrogen (NO) species was determined on peritoneal macrophages obtained from male BALB/c mice. Both *in vivo* and *in vitro* experiments were carried out, and propolis stimulation resulted to induce a slight augmentation in H_2O_2 releasing and a moderate inhibition of NO generation in a dose-dependent manner, suggesting that propolis acts on host non-specific immunity by macrophage activation [153]. Moreover, the effects of BGP on fungicidal activity of macrophages against *Paracoccidioides brasiliensis* were evaluated. Considering that cell-mediated immunity plays a significant role in host defence against this pathogen, peritoneal macrophages from BALB/c mice were stimulated with BGP and subsequently challenged with *P. brasiliensis*. This study suggested an increase in fungicidal activity of macrophages after propolis stimulation [151].

The role of toll-like receptors (TLR) in microbial pattern recognition and as well as the action of pro-inflammatory cytokines are important to trigger the initial events of immune response. Thus, the immunomodulatory action of BGP has been evaluated in regard to those mechanisms of innate immunity, and BGP treatment resulted to upregulate TLR-2 and TLR-4 expression, together with the production of pro-inflammatory cytokines (IL-1β and IL-6) in peritoneal macrophages and spleen cells of Male BALB/c mice treated with BGP of Sao Paulo State (200 mg/kg) for 3 days, suggesting the favourable action of BGP by enhancing immune responses in macrophages and spleen cells of treated mice [154]. Moreover, the action of BGP on antibody production was studied in rats immunized with bovine serum albumin, wherein BGP (10%), independently of the season of the year, stimulated antibody production, which was concluded to be a consequence of synergic effects, since isolated compounds and *B. dracunculifolia* exudates did not induced the antibody production [155].

Other studies in regard to the effects of BGP on immune response of acutely and chronically challenged stressed mice have been also investigated. After 3 days treatment under acutely stress conditions, BGP (200 mg/kg) restored TLR-2 and TLR-4 expression and increased IL-4 production in mice, in comparison with control [156, 157]. Meanwhile, BGP (200 mg/kg) during 7 days treatment in stressed mice increased the production of H_2O_2 in macrophages and reduced the alterations found in spleen [158]. Moreover, the BGP treatment (200 mg/kg) for 14 days in mice submitted to chronic stress exerted similar effects, demonstrating these experiments the immunomodulatory activity of BGP under *in vivo* experimental stress conditions [28, 152, 159, 160].

Once the effect of BGP on the initial steps of the immune response was evaluated in murine models, the immunomodulatory effect of BGP was evaluated on receptors expression, cytokine production and fungicidal activity of human monocytes. The BGP treatment (5, 10, 25, 50 and 100 μg/mL) resulted to upregulate TLR-4 and CD80 expression and decreased the production of TNF-α and IL-10 as concentration treatment increased. The fungicidal activity of human monocytes after incubation with BGP and challenged with *C. albicans* was increased in a dose-dependent manner. Moreover, cytokine production was reduced by blocking TLR-4, whereas the fungicidal activity was affected by blocking TLR-2, suggesting the involvement of these receptors in the mechanism of action of BGP [161].

Later, in an attempt to investigate the BGP constituents involved in the immunomodulatory effects on the innate immunity, caffeic and cinnamic acids

(5, 10, 25, 50 and 100 μg/mL) have been assessed on human monocytes. Caffeic acid downregulated TLR-2 and HLA-DR expression. Otherwise, cinnamic acid downregulated TLR-2, HLA-DR and CD80; meanwhile, it upregulated TLR-4 expression depending on concentration. Both phenolic acids inhibited TNF-α and IL-10 production, whereas they increased the fungicidal activity of monocytes against *C. albicans*, without affecting cell viability. These data suggested that caffeic and cinnamic acid are partially involved in BGP effect on cell receptors expression and cytokine production, although the fungicidal activity of monocytes treated with those phenolic acids could be due to different mechanisms, possibly involving reactive oxygen and nitrogen species [162, 163].

In another study, the antioxidant and anti-inflammatory action of BGP and caffeic acid were determined. By DPPH assay, FRS activity of caffeic acid (EC_{50}: 2.5 μg/mL) resulted to be more effective than that of BGP (EC_{50}: 18.51 μg/mL), and the treatment with BGP and caffeic acid (5, 10, 25, 50 and 100 μg/mL) exerted anti-inflammatory action, by inhibiting nitric oxide (NO) production in RAW 264.7 cells. Moreover, both treatments suppressed LPS-induced signalling pathways, namely p38 MAPK, JNK1/2 and NF-κB, and did not induce hepatotoxicity at the tested concentrations, suggesting that caffeic acid may be involved in the antioxidant and anti-inflammatory effects of BGP [164]. BGP and its main botanical source, *B. dracunculifolia* exudates have been demonstrated to exert cytotoxic effect on several human cancer cell lines, including HEp-2, CaCo2, HCT116, HT-29 and SW480, and moreover on canine osteosarcome cells [165–168].

The diverse phytochemical patterns found in samples of Brazilian propolis are a direct result of the country's vast plant biodiversity and the various geographical origins from where the samples were collected. Hence, it is imperative to conduct a more in-depth analysis of the chemical makeup of samples generated in various regions to ascertain the botanical source of the bioactive substances found in propolis throughout Brazil. The extensive pharmacological activity examined for Brazilian green, red, and brown propolis exemplify the significant biological interest in this beehive product, establishing it as one of the most extensively studied propolis varieties globally. Continuing research on Brazilian propolis is crucial for gaining a more accurate understanding of the potential applications of propolis from this biodiverse country, which includes both tropical and temperate varieties.

3.6. Uruguay

Uruguay is located within the temperate sub-humid grasslands biome, which stretches throughout the extensive plains of Southern South America, encompassing parts of Northeast and Central Argentina as well as Southern Brazil [169]. A study conducted on Uruguayan propolis revealed the presence of 22 distinct phenolic acids and flavonoids in six samples obtained from various locations of the country.

The presence of these constituents was detected by HPLC–PDA–UV, and resins from *Eucalyptus globulus*, *Populus* sp., *Betula* sp. and *Salix* sp. were suggested to be the botanical source of Uruguayan propolis. Those six propolis samples showed antibacterial activity against *B. subtilis* and *S. aureus* (MIC: 80–130 µg/mL); otherwise, the inhibitory effect of Uruguayan propolis against *E. coli* (MIC: 800–1000 µg/mL) was less efficient. Alkylperoxyl radical (ROO●) scavenging potential activity was additionally exhibited by those propolis samples (3.4–4.1 µg/mL) [170].

In another study, the chemical constitution of propolis collected from Montevideo was further investigated using HPLC–MS and NMR techniques and led to the isolation of eighteen flavonoids, four aromatic carboxylic acids and eleven phenolic acid esters, in addition to 3 new compounds elucidated: pinobanksin-3-*O*-isobutyrate and pinobanksin-3-*O*-(2-methyl)butyrate and 2-methyl-2-butenyl ferulate. The major constituents in that sample were pinobanksin-3-*O*-propanoate, pinobanksin-3-*O*-acetate, pinobanksin, pinostrobin, pinocembrin, techtochrysin, chrysin, galangin and cinnamyl *p*-coumarate, which suggested similarities to Southern Brazil, North American and European propolis [171]. Moreover, by RP–HPTLC, GC–MS and RP–HPLC analyses, it was determined that the phytochemical profile of propolis from Southern Brazil, Argentina and Uruguay was similar and correlated with that of *P. alba* resins; therefore, it was concluded that resins of poplar trees are the main plant origin of Uruguayan propolis [147].

Later on, a propolis sample from Montevideo was added to a study comparing it with samples from various locations. It was found that the propolis from Uruguay had similar qualities to propolis from the United States, New Zealand, Hungary, China, and other temperate regions that were included in the study. Furthermore, the propolis from Uruguay exhibited a moderate level of antioxidant activity as determined by the FRS method. Specifically, it had an antioxidant activity of at least 30% at a concentration of 20.0 µg/mL in the DPPH assay, and at least 45% at a concentration of 10.0 µg/mL in the β-carotene-linoleic acid system [82]. The antioxidant activities of 10 propolis samples obtained from Southern Uruguay were further assessed using in vitro methods such as FRS by ORAC and suppression of lipid and protein oxidation, as well as cellular tests. The Uruguayan propolis exhibited a significant level of FRS activity, as determined by the ORAC assay, with a value of 800 µmol trolox equiv/g propolis. The propolis from Uruguay demonstrated the ability to prevent the oxidation of LDL lipoproteins and the nitration of proteins. It achieved this by increasing the expression of endothelial nitric oxide synthase (eNOS) and inhibiting endothelial NADPH oxidase at the cellular level. These findings suggest that the propolis may have the potential to enhance the availability of nitric oxide in the endothelium, providing a possible benefit.

The volatile chemicals of three samples from Central-southern Uruguay were analyzed using the static headspace technique in combination with GC-MS. The results were then compared with the volatile content of propolis from Brazil, Estonia, and China. The primary components of the volatile profile of Uruguayan and Brazilian propolis were α-pinene and β-pinene (64.6–77.6%), which are both monoterpenes. The Brazilian propolis was characterized by a significant concentration of β-methyl crotonaldehyde (10.1%), whereas one of the samples from Uruguay included limonene (15.6%). The principal component analysis of the volatile profile in the samples revealed significant variations, resulting in distinct clusters. The study does not provide specific information about the geographical origin of Uruguayan and Brazilian propolis. However, based on the results obtained, it can be inferred that Brazilian propolis likely originates from the southern region. This inference is supported by the similarities in volatile composition between Brazilian propolis and Uruguayan propolis [173]. The findings from these investigations on Uruguayan propolis validate the chemical properties of propolis found in temperate regions.

3.7. Argentina

Argentina has a vast territory and a great diversity of ecological regions, including subtropical rain forests and temperate sub-humid grasslands or "Pampas." However, two thirds of Argentinean mainland is comprised by extensive arid and semi-arid region between the Andes and cold subantartic zones. Natural arid and semi-arid plain regions include western Chaco, Monte and Patagonian steppe, and these are ecologically similar to those semi-arid regions of North America, present in Sonora, Sinaloa and Great Basin, respectively [174, 175]. Several studies have differentiated propolis into two main groups, one from temperate zones and the other from tropical regions; however, few studies have been done in regard to propolis from arid and semi-arid areas [89]. Propolis from different regions of Argentina has been investigated, twenty-five samples collected in temperate, arid and semi-arid lands in Northern Argentina (Santiago del Estero, Tucumán, Chaco, Salta, Catamarca, Jujuy and Misiones) were chemically and biologically analysed. By absorption spectra, RP–HPTLC and RP–HPLC analysis, it was found that 16 of the 25 samples presented a similar phenolic profile to that exhibited by samples from Southern Brazil and Uruguay. The main phenolic constituents identified in those 16 samples were pinobanksin, pinocembrin, chrysin, galangin, tectochrysin and 1,1-dimethylallylcaffeic acid [176]. Propolis samples from Catamarca and Tucumán presented the highest phenolic content, exhibiting greater amounts of pinocembrin, 1,1-dimethylallylcaffeic acid, ferulic acid and cinnamic acid in comparison with the other 20 samples analyzed by Isla et al. [176]. Additionally, these two propolis presented the higher antibacterial activity against *S. aureus*

ATCC 25923 As well (inhibition zone: 4 and 5 mm, respectively) and presented the high FRS activity by DPPH assay (>40%) [176].

Moreover, other biological properties of propolis from Tucumán have been tested, including cytoxicity with the lethality test of *A. salina* (LD_{50}: 100 μg/mL), toxicity and mutagenicity on both *S. typhimurium* TA98 and TA100 (did not exhibit toxicity at 300 μg/plate), genotoxicity on *Allium cepa* (not presented) and antimutagenicity, inhibiting the effect of isoquinoline (IQ) and 4-nitro-o-phenylenediamine (NPD) (ID_{50}: 40 and 20 μg/plate, respectively). Furthermore, the chemical constituent 2′,4′-dihydroxychalcone of Tucumán propolis showed cytotoxic activity (LC_{50}: 0.5 μg/mL) and was able to inhibit the mutagenicity of IQ (ID_{50}: 1 μg/plate), whereas genotoxic or mutagenic effects were not observed [177]. Additionally, the antimycotic activity of Tucumán propolis and its compounds, pinocembrin and galangin, has been tested against *Trichoderma* spp., *Penicillium notatum*, *Aspergillus niger*, *Fusarium* sp., *Phomopsis* spp., *Saccharomyces carlsbergensis*, *Rhodotorula* spp. by bioautography, hyphal radial growth, hyphal extent and microdilution in liquid medium. Wherein propolis inhibited filamentous fungal growth and exhibited an antifungal moderate activity (MIC: 77–349 μg/mL). Pinocembrin, and galangin presented higher antifungal activity (MIC: 77 349 μg/mL), and their assumed as partially responsible for the fungitoxic activity of Tucumán propolis [178].

Similar results have been obtained from Tucumán propolis against human opportunistic and pathogenic fungi, specifically on dermatophytes and yeasts (*C. albicans* ATCC 10231, *C. tropicalis* C 131, *S. cerevisiae* ATCC 9763, *C. neoformans* ATCC 32264, *A. flavus* ATCC 9170, *A. fumigatus* ATTC26934, *A. niger* ATCC 9029, *T. rubrum* C 110, *T. mentagrophytes* ATCC 9972, and *M. gypseum* C 115). All the dermatophytes and yeasts were inhibited by three different Tucumán propolis samples (MIC: 16–125 μg/mL). The most susceptible fungi were *M. gypseum*, *T. mentagrophytes* and *T. rubrum*. A bioassay-guided isolation of Tucumán propolis and chemical characterization by NMR and HPLC–ESI-MS/MS yield two bioactive chalcones: 2′,4′-dihydroxy-3′-methoxychalcone and 2′,4′dihydroxychalcone that displayed strong activity against clinical isolates of *T. rubrum* and *T. mentagrophytes* (MIC: 1.9–2.9 μg/mL). In addition, pinocembrin, galangin, 7-hydroxy-8-methoxyflavanone presented a moderate antifungal activity. By the presence of those compounds was identified in exudates of *Zuccagnia punctata* Cav. (Caesalpinieae), which was assigned as the botanical origin from Tucumán propolis samples [179].

Other studies focused on the characterization of propolis from Catamarca Province in Northwestern Argentina have been done. Propolis collected from El rincón presented FRS activity by ABTS assay (SC_{50}: 6.9 μg/mL) and by β-carotene-linoleic acid antioxidant assay (SC_{50}: 2.0 μg/mL), additionally exhibited antibacterial effect on methicillin resistant *S. aureus* (MRSA) by the microdilution method and bioautographic assays (MIC: 65 µg/mL). This propolis led to the isolation and characterization by NMR of twelve compounds, wherein the most bioactives were 2′,4′-dihydroxy-3′-methoxychalcone, 2′,4′-dihydroxychalcone, 2′,4′,4-trihydroxy-6′-methoxychalcone, 5-hydroxy-4′,7′-dimethoxyflavone, 4′,5′-dihydroxy-3,7,8-trimethoxyflavone and 7-hydroxy-5,8-dimethoxyflavone [180].

FRS activity by DPPH assays of Argentinean propolis collected in the regions of Mendoza, Rio Negro, La Pampa, and Entre Rios was evaluated. Almost all of the propolis samples presented FRS activity (40-60% at 20 µg/mL), in exception to the samples from La Pampa and Entre Rios. Greater amounts of caffeic acid, ferulic acid, caffeic acid phenetyl ester were found by HPLC–PDA–UV analysis in propolis samples with the stronger FRS activity. In general, the presence of flavonoids, such pinocembrin, chrysin, pinobanksin, pinobanksin-3-O-acetate and galangin, was found in almost all samples [181]. Moreover, three samples collected from two different pythogeographical regions (Prepuna and Monte) exhibited FRS activity by DPPH assays (IC_{50}: 28–43 µg/mL) and by β-carotene–linoleic acid antioxidant assay (IC_{50}: 2.0–8.4 µg/mL, respectively). The samples from Monte region showed the highest inhibitory effect on different strains of *S. aureus* and *E. faecalis* (MIC_{100}: 50 and 100 µg/mL, respectively). Nine compounds were identified by HPLC–DAD–UV, and two of them were only identified in samples from Monte region 2′4′-dihydroxychalcone and 2′,4′-dihydroxy 3′-methoxychalcone. These two bioactive chalcones, present in propolis from El Rincon and Monte region, have been detected in propolis from Tucumán and in *Zuccagnia punctata*, a perennial shrub in Argentinean arid regions, which is identified as botanical source of this propolis from Argentinean arid lands [182].

The chemical composition of propolis collected from the Andean locality of Bauchaceta in San Juan province was characterized by HPLC–ESI-MS/MS and GC–MS techniques and revealed a lignan and volatile organic acid profile that matched with the exudates of *Larrea nitida*, which was suggested as its main botanical source. Andean propolis presented antifungal activity against Dermatophytes and yeasts (MIC: 31.3–125 µg/mL). A bioassay-guided isolation of the most active compounds yield two lignans characterized as 3′-methyl-nordihydroguaiaretic acid (NDGA) and nordihydroguaiaretic acid (NDGA) that showed high inhibitory activity against *T. mentagrophytes*, *T. rubrum*, *M. gypseum* (15.6–31.25 µg/mL) and clinical isolates of *Candida* spp., *Cryptococcus* spp., *T. rubrum* and *T. mentagrophytes* (MIC: 31.3 62.5 µg/mL). In addition, 4-[4-(4-hydroxy-phenyl)-2,3-dimethyl-butyl]-benzene-1,2-diol, 4′-methyl-nordinydroguaiaretic acid and two epoxylignans meso-(rel 7S,8S,7′R,8′R)-3,4,3′,4′-tetrahydroxy-7,7′-epoxylignan and (7S,8S,7′S,8′S)-3,3′,4′-trihydroxy-4-methoxy-7,7′-epoxylignan) were isolated and identified by their spectroscopic data in NMR experiments. These six compounds were isolated from propolis for the first time [183]. Similar antifungal activity results were obtained from other eleven samples collected in San Juan province, and the identification and isolation of MNDGA and NDGA. The flavonoids chrysin, pinocembrin and galangin were the most bioactive compounds [184].

A reverse phase LC–DAD–MS method has been developed to quantify phenolic acids and flavonoids present in propolis and found that the most abundant constituents from Argentinean, European and Chinese propolis were chrysin (2–4%), pinocembrin (2–4%), pinobanksin-acetate (1.6–3%) and galangin (1–2%) [185]. Furthermore, the typical fingerprint of propolis from Argentina, Italy and Spain was determined by on-line HPLC–ESI/MS analysis, and those propolis samples showed the same total ion chromatogram (TIC) profile, in addition to a similar amount of pinocembrin (39-49% of the total identified flavonoids), suggesting a poplar-type origin for Argentinean propolis [186]. Interestingly, the subtropical montane forest of "El Siambón" in Tucumán region is characterized by the presence of native vegetation including members of Lauraceae, Myrtaceae, Fabaceae, Juglandaceae, Salicaceae and Nyctaginaceae families, as well as the introduced poplar, eucalyptus and pinus trees, which are described as the most visited trees by honeybees for resins in the Tucuman region [178].

However, in another study, propolis samples from 30 different regions of Santiago del Estero province were analysed and classified into three groups according to the main arboreal species in the region. Nineteen chemical constituents were identified in these 30 samples, pinocembrin, quercetin, kaempferol, chrysin, salicylic, 4-hydroxybenzoic, benzoic, ferulic and gallic acids were observed in most of the analysed samples. Some quantitative differences of each component were found in the three groups, and interestingly, no *Populus* species were found in apiary environments, where the most abundant plant species were *Geoffroea decorticans*, *Prosopis alba*, *Prosopis nigra*, *Schinopsis lorentzii*, *Acacia aroma*, *Cercidium praecox*, *Schinus fasciculatus*, and *Larrea divaricata*. These studies suggest that flavonoids characteristic of temperate zones, which are present in Argentinean propolis, could be gathered from other plant species than poplar [187].

Furthermore, the antibacterial, antiradical and antioxidant activities of those 30 propolis samples from Santiago del Estero province were analysed, and variability in bioactivity was found. All samples presented a substantially similar inhibitory effect on *S. aureus* ATCC 25923 (8.5–11.4 mm), and about 77% of the samples showed an inhibitory zone with diameter longer than 9 mm. Additionally, there was observed a moderate correlation between antibacterial activity and total polyphenol and flavonoid contents. However, antimicrobial activity correlated better with pinocembrin content than with total polyphenol content. Results for the antioxidant activity by the β-carotene–linoleic acid assay presented high variability among samples (16.2–84.7%) as consequence of quantitative differences in chemical composition. Similar results were obtained by FRS activity on DPPH assay (20.4–89.9%) [188].

Studies conducted on Argentinean propolis have shown that samples collected from various ecological regions in the country have different chemical profiles. However, most of the analyzed samples contain flavonoids, and compounds like lignans and chalcones can be used as markers to indicate the presence of specific botanical sources. Therefore, the diverse phytochemical composition of Argentinean propolis can be attributed to several plants, including Populus alba, Larrea nitida, Larrea divaricata, and Zuccagnia punctata.

3.8. Chile

Chile stretches along the Southwestern coast of South America, bordered by the Andes mountain range to the East and the Pacific Ocean to the West. This geographical isolation has led to the development of a distinct and diverse flora, with numerous plant species that are found only in Chile. Chile possesses the Atacama Desert in the northern part of the country, a Mediterranean-type region in the north-central area, temperate woods in the south-central region, and subpolar forests in the southern part [189, 190].

Some studies have been carried out in Chilean propolis collected from different regions. A propolis sample from Quebrada Yaquil in Santa Cruz (Region VI) in the Mediterranean semi-arid region of central Chile have led to the isolation and structural determination by NMR of five lignans, including 3 novel compounds: 1-(4-hydroxy-3-methoxyphenyl)-1,2-bis{4-[(E)-3-acetoxypropen-1-yl]-2-methoxyphenoxy}propan-3-ol acetate, and two different optical isomers ($[\alpha]^{25}D$ + 8.8° and $[\alpha]^{25}D$–15.6°, respectively) of 1-(4-hydroxy-3-methoxyphenyl)-2-{4-[(E)-3-acetoxypropen-1-yl]-2-methoxyphenoxy}propan-1,3-diol 3-acetate, in addition to the already reported 3-acetoxymethyl-5-[(E)-2-formylethen-1-yl]-2-(4-hydroxy-3-methoxyphenyl)-7-methoxy-2,3-dihydrobenzofuran, and 3-acetoxymethyl-5-[(E)-3-acetoxypropen-1-yl)]-2-(4-hydroxy-3-methoxyphenyl)-7-methoxy-2,3-dihydrobenzofuran. The hives were located in a region with the dominant species sclerophyllous shrubs and herbaceous species, such as *Lithrea caustica*, *Quillaja saponaria*, *Cryptocarya alba*, *Kageneckia oblonga*, *Colliguaja odorifera*, *Trevoa trinervis*, *Baccharis linearis*, *Peumus boldus*, *Madia sativa*, *Helenium aromaticum* and *Pasithea caerulea*, which were considered as possible botanical sources of resins [190].

In another study, propolis was collected from hives located at Cuncumen (also Mediterranean-type climate), and fractionated to yield viscidone, vanillin, 3′,4′-(methylendioxy)acetophenone, 3-ethoxy-4-methoxybenzaldehyde, cinnamic acid and 3-methoxy-4-hydroxymethyl ester. These chemical compounds were already described; however, this was the first report on propolis composition of an arid, and a Mediterranean-type climate area. Additionally, the frequency of pollen grains observed in this Cuncumen propolis by optical and scanning electronic microscopy (SEM) allowed to suggest that resins of *Eucalyptus* spp. and *Salix humboldtiana* are the main botanical source of *A. mellifera* in central Chile. Pollen grains of *B. linearis*, *Q. saponaria* and *L. caustica* were also present in propolis sample in lower amounts [191]. Therefore, in order to understand more about the chemical composition and botanical sources available for honeybees in Central Chile, the same research group studied other sample from sclerophyllous shrubland coast (Colliguay), and by chromatographic procedures led to the isolation and characterization by NMR of seven phenolic compounds, including pinocembrin, acacetin, galangin, izalpin, kaempferide, prenyletin and diarytheptane. The botanical origin of Colliguay propolis was investigated by palynological analysis in optic microscopy, and the most abundant pollen grains and leaf fragments found in the sample were related to the plants *Escallonia pulverulenta*, *S. humboldtiana* and *Eucalyptus globulus*, which suggested their participation as botanical sources of this Chilean propolis. Probably honeybees obtain resins rich in pinocembrin from *E. pulverulenta*, since this flavanone was formerly reported from this source [192].

Other studies have been carried out in propolis from Central Chile. Six samples collected from the region of Santiago (Curacaví, Lo Cañas, Buin, Caleu, Cajón del Maipo and pireque) were analyzed by HPLC–ESI-MS/MS, and 30 chemical phenolic compounds were identified, including pinocembrin, pinobanksin, pinobanksin-3-*O*-alkanoates, caffeic acid phenethyl ester (CAPE), chrysin, galangin, kaempferol, kaempferide, ferulic acid, quercetin and quercetin methyl ether derivatives. The antioxidant properties of those samples were analyzed by FRAP, ORAC-FL, ORAC-PGR and DPPH radical methods. All samples exhibited a different FRS activity. The samples that presented CAPE (Curacaví, Buin and Cajón del Maipo) and quercetin (Buin and Lo Cañas) exerted the best antioxidant activity. Pinobanksin was found in all the samples, a compound that would be a suitable candidate for the standardization of propolis from the region of Santiago [193]. In a subsequent study, the anti-inflammatory effect of propolis from Caleu and Buin was tested through mice ear edema model, in addition to the *in vitro* assessment of nitric oxide (NO) production by RAW 264.7 cells stimulated with lipopolysaccharide (LPS). Buin propolis presented anti-inflammatory effect in the murine model (64%), and moreover, significantly decreased the NO release in RAW 264.7 cells in a dose-dependent manner [194].

In addition, a propolis sample collected from San Vicente de Tagua-Tagua (VI Region) was biologically assessed, and showed FRS activity by DPPH (100% at 80 μg/mL) and scavenger effect on superoxide anion (100% at 0.78 μg/mL), as well this Chilean propolis presented antiproliferative activity (at 80 μg/mL for 72 h incubation) by MTT assays on human tumour cell lines KB, Caco-2 and DU-145 (

9, 45 and 23% cell viability, respectively). Galangin, caffeic acid, p-coumaric acid, ferulic acid and CAPE were identified by HPLC analysis in San Vicente sample, and since the most abundant plant species in this region were *Peumus boldus, Q. saponaria*, *P. alba* and *Pinus radiata*, thc botanical origin of this propolis according to the presence of those compounds would be the resins of *P. alba* [195].

Moreover, the chemical constitution, the botanical origin and antibacterial activity of twenty samples collected from Central and Southern regions of Chile (Valparaíso, Metropolitana, Libertador Bernardo O'Higgins and La Araucanía Regions) were investigated. Quercetin, myricetin, kaempferol, pinocembrin, coumaric acid, caffeic acid and CAPE were identified in propolis samples by HPLC–MS. All Chilean propolis samples presented a growth inhibitory effect on *S. mutans* and *S. sobrinus* (MIC: 0.9–8.2 μg/mL). By palynological analysis, the plant structures of native species, such as *Trevoa quinquenervia*, *Aristotelia chilensis*, *L. caustica*, *Retanilla trinervia*, *Q. saponaria*, and species of the genus *Escallonia* were found in propolis samples from central regions, whereas *Lotus uliginosus*, *Aextoxicon punctatum*, *B. linearis* and *Eucryphia cordifolia* were identified in samples from Southern Chile, additionally no structures of the genus *Populus* were detected in all the samples. These results suggested that honeybees could obtain CAPE and those flavonoids from other species rather than poplars [196].

With the aim to determine whether the bioactivity against *S. mutans* of Chilean propolis from "La araucanía" region was influenced by the year of collection, three different samples from spring of 2008, 2010 and 2011 were studied. The chemical composition of the annual samples presented qualitative differences by LC-MS analysis. Apigenin, genistein, kaempferol, myricetin, pinocembrin, quercetin, CAPE, caffeic acid, *p*-coumaric acid and ferulic acid were found in all the samples. Otherwise, the presence of daidzein, rutine, and chlorogenic and gallic acid was not constant. The antimicrobial activity of the annual samples did not presented significant variations (MIC: 0.91, 0.22 and 0.39 μg/mL); however, the biofilm formation in *S. mutans* cultures treated with Chilean propolis showed to be influenced by the year of collection [197].

These reports provide important information about chemical composition of Chilean propolis. At present, two main types of Chilean propolis are described, those with lignans as main compounds, and the others with flavonoids and phenolic acids related to propolis collected from temperate regions; however, the palynological evidence obtained indicated that native plants, such as *S. humboldtiana*, *E. globulus*, *E. pulverulenta*, *L. uliginosus*, *B. linearis*, *Q. saponaria* and *L. caustica* are related to the botanical origin of Central and Southern Chile. Moreover, the antibacterial, cytotoxic, antiproliferative and antioxidant activities of Chilean propolis make it a bioactive product for further investigations.

4. CONCISE SUMMARY

The propolis generated by honeybees in the Americas is a significant reservoir of several bioactive chemicals, including nemorosone and other prenylated benzophenones, artepillin C, CAPE, labdane-type terpenes, chalcones, flavonoids, lignans, and more. Various plants in different ecological zones around the continent produce these secondary metabolites through biosynthesis. Propolis from the Americas has unique characteristics, with certain samples including a combination of temperate and tropical chemical composition. Additionally, certain mixes of compounds can be found in propolis from both North and South America, resulting in a wide range of propolis diversity. Furthermore, propolis obtained from temperate regions is not limited to being derived solely from poplar, birch, pines, and horse chestnut exudates. Recent research has discovered that propolis from the United States and Honduras is sourced from liquidambar species, offering a unique and transcontinental alternative source of resinous material.

Currently, the chemical composition of propolis from North America and South America is only partially understood. However, further research is necessary to fully comprehend and identify the bioactive compounds present in these samples, as well as to determine the specific mechanisms by which they exert their pharmacological effects. Furthermore, further research is needed to comprehend the foraging behavior of plant resin bees, the function of propolis within the hive, and the advantages of botanical chemistry found in the various ecological regions of America. This necessitates interdisciplinary collaboration to establish a connection between bee health and the implicated ecosystem. Given the wide range of biological activities, the extensive variability, and the intricate nature of North American and South American propolis, it is imperative to establish a more accurate classification system for this diverse propolis. This classification should incorporate both the qualitative and quantitative analysis of the plants from which it originates, as well as the biological properties of this product from beehives.

REFERENCES

1. Bankova V. Recent trends and important developments in propolis research. eCAM. 2005;2(1):29–32. doi:10.1093/ecam/neh059
2. Grimaldi D, Engel MS. Evolution of the insects. 1st ed. Cambridge: Cambridge University Press; 2005. 772 p. doi:978-0-521-82149-0
3. Michener CD. The bees of the world. 2nd ed. Maryland: The Johns Hopkins University Press; 2007. 992 p. doi:978-0-8018-8573-0
4. Wilson MB, Spivak M, Hegeman AD, Rendah, A, Cohen JD (10). Metabolomics reveals the origins of antimicrobial plant resins collected by honey bees. Plos One. 2013;8(10):e77512. doi:10.1371/journal.pone.0077512

5. Winston ML, Michener CD. Dual origin of highly social behavior among bees. Proc Natl Acad Sci USA. 1977;74(3):1135–1137.

6. Cortopassi-Laurino M, Imperatriz-Fonseca VL, Roubik DW, Dollin A, Heard, T, Aguilar I, Venturieri G, Eardley C, Nogueira-Neto P. Global meliponiculture: challenges and opportunities. Apidologie. 2006;37(2006):275–292. doi:10.1051/apido:2006027

7. Han F, Wallberg A, Webster MT. From where did the Western honeybee (Apis mellifera) originate? Ecol Evol. 2012;2(8):1949–1957. doi:10.1002/ece3.312

8. Pechhacker H, Joshi SR, Chatt A, Horn H. Beekeeping around the world. Bee World. 2001;82(2):99–109. doi:10.1080/0005772X.2001.11099511

9. vanEngelsdorp D, Meixner MD. A historical review of managed honey bee population in Europe and the United States and the factors that may affect them. J Invertebr Pathol. 2010;103(2010):S80–S95. doi:10.1016/j.jip.2009.06.011

10. Temaru E, Shimura S, Amano K, Karasawa T. Antibacterial activity of honey from stingless honeybees (Hymenoptera; Apidae; Meliponinae). Pol J Microbiol. 2007;56(4):281–285.

11. Michener CD. The Meliponini. In: Vit P, Pedro SRM, Roubik, D, editors. Pot-honey: a legacy of stingless bees. 1st ed. New York: Springer Sicence; 2013. pp. 3–18. doi:10.1007/978-1-4614-4960-7

12. Simone-Finstrom MD, Spivak M. Increased resin collection after parasite challenge: a case of self-medication in honey bees? Plos One. 2013;7(3):e34601. doi:10.1371/journal.pone.003460

13. Wilson MB, Brinkman D, Spivak M, Gardner G, Cohen JD. Regional variation in composition and antimicrobial activity of US propolis against Paenibacillus larvae and Ascosphaera apis. J Invertebr Pathol. 2015;124(2015):44–50. doi:10.1016/j.jip.2014.10.005

14. Bankova V, de Castro SL, Marcucci MC. Propolis: recent advances in chemistry and plant origin. Apidologie. 2000;31(2000):3–15

15. Castaldo S, Capasso F. Propolis, an old remedy used in modern medicine. Fitoterapia. 2002;73(Suppl. 1):S1–S6

16. Ghisalberti EL. Propolis: a review. Bee World. 1979;60:59–84

17. Kuropatnicki AK, Szliszka E, Krol W. Historical aspects of propolis research in modern times. eCAM. 2013;2013:964149. doi:10.1155/2013/964149

18. Marcucci MC. Propolis: chemical composition, biological properties and therapeutic activity. Apidologie. 1995;26:83–99.

19. Greenaway W, Scaysbrook T, Whatley FR. The analysis of bud exudate of Populus x euramericana, and of propolis, by gas chromatography-mass spectrometry. Proc R Soc Lond B. 1987;232:249–272.
20. Gençay-Çelemli O. Chemical properties of propolis collected by stingless bees. In: Vit P, Pedro SRM, Roubik, D, editors. Pot-Honey: A legacy of stingless bees. 1st ed. New York: Springer Sicence; 2013. pp. 525–537. doi:10.1007/978-1-4614-4960-7
21. Bankova V, Popova M. Propolis of stingless bees: a promising source of biologically active compounds. Pharmacogn Rev. 20074;1(1):88–92. doi:10.1186/1752-153X-8-28
22. Freitas AS, Vit P, Barth OM. Pollen analysis of geopropolis and propolis from stingless bees. In: Vit P, Roubik DW, editors. Stingless bees process honey and pollen in cerumen pots. 1st ed. Facultad de Farmacia y Bioanálisis, Mérida, Venezuela: Universidad de Los Andes; 2013. pp. 1–7. doi:978-980-11-15
23. Banskota AH, Tezuka Y, Prasain JK, Matsushige K, Saiki I, Kadota S. Chemical constituents of Brazilian propolis and their cytotoxic activities. J Nat Prod. 1998;61(7):896–900. doi:10.1021/np980028c
24. Pereira AS, Norsell M, Cardoso JN, Aquino-Neto JR, Ramos MF. Rapid screening of polar compounds in Brazilian propolis by high-temperature high-resolution gas chromatography-mass spectrometry. J Agric Food Chem. 2000;48(11):5226–5230. doi:10.1021/jf000594p
25. Freitas MO, Ponte FAF, Lima, MAS, Silveria ER. Flavonoids and triterpenes from the nest of the stingless bee Trigona spinipes. J Braz Chem Soc. 2008;19(3):532–535. doi:10.1590/S0103-50532008000300022
26. Vit P, Medina M, Enríquez ME. Quality standards for medicinal uses of Meliponinae honey in Guatemala, Mexico and Venezuela. Bee World. 2004;85(1):2–5. doi:10.1080/0005772X.2004.11099603
27. Salatino A, Fernandes-Silva CC, Abbud Righi A, Salatino MLF. Propolis research and the chemistry of plant products. Nat Prod Rep. 2011;28:925–936. doi:10.1039/C0NP00072H
28. Sforcin JM, Banjova V. Propolis: is there a potential for the development of new drugs? J Ethnopharmacol. 2011;133(2):253–260. doi:10.1016/j.jep.2010.10.032
29. Liberio S, Pereira AL, Dutra R, Reis AS, Araujo MJ, Mattar NS, Silva L, Ribeiro MN, Nascimento F, Guerra RN, Monteiro-Neto V. Antimicrobial activity against oral pathogens and immunomodulatory effects and toxicity of geopropolis produced by the stingless bee Melipona fasciculata Smith. BMC Complement Altern Med. 2011;11:108. doi:10.1186/1472-6882-11-108

30. Sforcin JM, Fernandes A, Lopes CAM, Bankova V, Funari SRC, 2000, 73, 243. Seasonal effect on Brazilian propolis antibacterial activity. J Ethnopharmacol. 2000;73(1–2):243–249. doi:10.1016/S0378-8741(00)00320-2
31. Velazquez C, Navarro M, Acosta A, Angulo A, Dominguez Z, Robles R, Robles-Zepeda R, Lugo E, Goycoolea FM, Velazquez EF, Astiazaran H, Hernandez J. Antibacterial and free-radical scavenging activities of Sonoran propolis. J Appl Microbiol. 2007;103(5):1747–1756. doi:10.1111/j.1365-2672.2007.03409.x
32. Popova, M, Silici S, Kaftanoglu O, Bankova V. Antibacterial activity of Turkish propolis and its qualitative and quantitative chemical composition. Phytomedicine. 2005;12(3):221–228. doi:10.1016/j.phymed.2003.09.007
33. Yang H, Dong Y, Du H, Shi H, Peng Y, Li X. Antioxidant compounds from propolis collected in Anhui, China. Molecules. 2011;16(4):3444–3455. doi:10.3390/molecules16043444
34. Alday-Provencio S, Diaz G, Rascon L, Quintero J, Alday E, Robles-Zepeda R, Garibay-Escobar A, Astiazaran H, Hernandez J, Velazquez C. Sonoran propolis and some of its chemical constituents inhibit in vitro growth of Giardia lamblia trophozoites. Planta Med. 2015;81(9):742–747. doi:10.1055/s-0035-1545982
35. Freitas SF, Shinohara L, Sforcin JM, Guimarães S. In vitro effects of propolis on Giardia duodenalis trophozoites. Phytomedicine. 2006;13(3):170–175. doi:10.1016/j.phymed.2004.07.008
36. da Silva-Cunha I, Salomão K, Shimizu M, Bankova V, Ramalho CA, Lisboa de Castro S, Marcucci M. Antitrypanosomal activity of Brazilian propolis from Apis mellifera. Chem Pharm Bull. 2004;52(2):602–604.
37. Machado GM, Leon LL, De Castro SL. Activity of Brazilian and Bulgarian propolis against different species of Leishmania. Mem Inst Oswaldo Cruz. 2007;102(1):73–77. doi:10.1590/S0074-02762007000100012
38. Salomão K, Souza EM, Henriquez-Pons A, Barbosa HS, de Castro SL. Brazilian green propolis: effects in vitro and in vivo on Trypanosoma cruzi. eCAM. 2011;10(2011):185918. doi:10.1093/ecam/nep014
39. Sforcin JM, Fernandes A, Lopes CAM, Funari SRC, Bankova V. Seasonal effect of Brazilian propolis on Candida albicans and Candida tropicalis. J Venom Anim Toxins. 2001;7(1):139–144. doi:10.1590/S0104-79302001000100009
40. Hegazi AG, El Hady FKA. Egyptian propolis: 3. Antioxidant, antimicrobial activities and chemical composition of propolis from reclaimed lands. Z Naturforsch C. 2002;57(3–4):395–402.
41. Majiene DT, Pavilonis S, Savickas A, Martirosyan DM. Antifungal and antibacterial activity of propolis. Curr Nutr Food Sci. 2007;3(4):304–308. doi:10.2174/157340131070304030 4

42. Búfalo MC, Figueiredo AS, de Sousa JP, Candeias JM, Bastos JK, Sforcin JM. Anti-poliovirus activity of Baccharis dracunculifolia and propolis by cell viability determination and real-time PCR. J Appl Microbiol. 2009;107(5):1669–1680. doi:10.1111/j.1365-2672.2009.04354.x

43. Paintz M, Metzner J. On the local anaesthetic action of propolis and some of its constituents. Pharmazie. 1979;34(12):839–841.

44. Paulino N, Dantas AP, Bankova V, Longhi DT, Scremin A, de Castro SL, Calixto JB. Bulgarian propolis induces analgesic and anti-inflammatory effects in mice and inhibits in vitro contraction of airway smooth muscle. J Pharmacol Sci. 2003;93(3):307–313.

45. Khayyal MT, el-Ghazaly MA, el-Khatib AS. Mechanisms involved in the antiinflammatory effect of propolis extract. Drugs Exp Clin Res. 1993;19(5):197–203.

46. Conti BJ, Santiago KB, Búfalo MC, Herrera YF, Alday E, Velazquez C, Hernandez J, Sforcin JM. Modulatory effects of propolis samples from Latin America (Brazil, Cuba and Mexico) on cytokine production by human monocytes. J Pharm Pharmacol. 2015;67(10):1431–1438. doi:10.1111/jphp.12431

47. Orsatti CL, Missima F, Pagliarone AC, Sforcin JM. Th1/Th2 cytokines' expression and production by propolis-treated mice. J Ethnopharmacol. 2010;129(3):314–318. doi:10.1016/j.jep.2010.03.030

48. Borrelli F, Izzo AA, Di Carlo G. Effect of a propolis extract and caffeic acid phenethyl ester on formation of aberrant crypt foci and tumors in the rat colon. Fitoterapia. 2002;73(S1):S38–S43. doi:10.1016/S0367-326X(02)00189-2

49. El-khawaga OAY, Salem TA, Elshal MF. Protective role of Egyptian propolis against tumor in mice. Clin Chim Acta. 2003;338(1–2):11–16. doi:10.1016/S0009-8981(03)00323-1

50. Sawicka D, Car H, Borawska MH, Nikliński J. The anticancer activity of propolis. Folia Histochem Cytobiol. 2012;50(1):25–37. doi:10.2478/18693

51. Hernandez J, Goycoolea FM, Quintero J, Acosta A, Castañeda M, Dominguez Z, Robles R, Vazquez-Moreno L, Velazquez EF, Astiazaran H, Lugo E, Velazquez C. Sonoran propolis: chemical composition and antiproliferative activity on cancer cell lines. Planta Med. 2007;73(14):1469–1474. doi:10.1055/s-2007-990244

52. Alday E, Valencia D, Carreño AL, Picerno P, Piccinelli AL, Rastrelli L, Robles-Zepeda R, Hernandez J, Velazquez C. Apoptotic induction by pinobanksin and some of its ester derivatives from Sonoran propolis in a B-cell lymphoma cell line. Chem Biol Interact. 2015;242:35–44. doi:10.1016/j.cbi.2015.09.013

53. Li F, Awale S, Tezuka Y, Kadota S. Cytotoxicity of constituents from Mexican propolis against a panel of six different cancer cell lines. Nat Prod Commun. 2010;5(10):1601–1606.

54. Li F, Awale S, Tezuka Y, Esumi H, Kadota S. Study on the constituents of Mexican propolis and their cytotoxic activity against PANC-1 human pancreatic cancer cells. J Nat Prod. 2010;73(4):623–627. doi:10.1021/np900772m

55. Bankova V. Chemical diversity of propolis and the problem of standardization. J Ethnopharmacol. 2005;100(1–2):114–117. doi:10.1016/j.jep.2005.05.004

56. Lotti C, Campo Fernandez M, Piccinelli AL, Cuesta-Rubio O, Márquez Hernández I, Rastrelli L. Chemical constituents of red Mexican propolis. J Agric Food Chem. 2010;58(4):2209–2213. doi:10.1021/jf100070w

57. Piccinelli AL, Campo Fernandez M, Cuesta-Rubio O, Márquez Hernández I, De Simone F, Rastrelli L. Isoflavonoids isolated from Cuban propolis. J Agric Food Chem. 2005;53(23):9010–9016. doi:10.1021/jf0518756

58. Cuesta-Rubio O, Piccinelli AL, Rastrelli L. Tropical propolis: recent advances in chemical components and botanical origin. In: Rai MK, Geoffrey R, Cordell GC, Martínez JL, Marinoff M, Rastrelli L, editors. Medicinal plants: Biodiversity and drugs. 1st ed. New York: CRC Press, Taylor & Francis Group; 2012. pp. 209–240. doi:10.1201/b12527-17

59. Bankova V, Boudourova-Krasteva G, Sforcin JM, Frete X, Kujumgiev A, Maimoni-Rodellab R, Popov, S. Phytochemical evidence for the plant origin of Brazilian propolis from São Paulo state. Z Naturforsch C. 1999;54(5–6):401–405.

60. Salatino A, Teixeira EW, Negri G, Message D. Origin and chemical variation of Brazilian propolis. eCAM. 2005;2(1):33–38. doi:10.1093/ecam/neh060

61. Silva-Carvalho R, Baltazar F, Almeida-Aguiar C. Propolis: a complex natural product with a plethora of biological activities that can be explored for drug development. eCAM. 2015;2015:206439. doi:10.1155/2015/206439

62. Burdock GA. Review of the biological properties and toxicity of bee propolis (propolis). Food Chem Toxicol. 1998;36(4):347–363. doi:10.1016/S0278-6915(97)00145-2

63. Ghedira K, Goetz P, Le Jeune R. Propolis. Phytothérapie. 2009;7:100–105. doi:10.1007/s10298-009-0377-8

64. Robles-Zepeda RE, Hernández-Martínez J, Garibay-Escobar A, Valencia-Rivera DE, Velázquez-Contreras CA. Botanical Origin and Biological Activity of Propolis. In: Rai MK, Geoffrey R, Cordell GC, Martínez JL, Marinoff M, Rastrelli L, editors. Medicinal plants: Biodiversity and drugs. 1st ed. New York: CRC Press, Taylor & Francis Group; 2012. p. 570–597. doi:10.1201/b12527-17

65. Sawaya ACHF, Cunha IBS, Marcucci MC, de Oliveira-Rodrigues RF, Nogueira-Eberlin M. Brazilian propolis of Tetragonisca angustula and Apis mellifera. Apidologie. 2006;37(3):398–407. doi:10.1051/apido:2006011

66. Sawaya ACHF, da Silva-Cunha IB, Marucci MC, Aidard DS, Almeida-Silva EC, Lopes-Carvalho CA, Nogueira-Eberlin M. Electrospray ionization mass spectrometry fingerprinting of propolis of native Brazilian stingless bees. Apidologie. 2007;38(1):93–103. doi:10.1051/apido:2006058

67. Daubenmire R. Plant geography: with special reference to North America. 1st ed. New York: Academic Press; 1978. 346 p. doi:10.1016/B978-0-12-204150-1.50006-0

68. Woodward FI, Lomas MR, Kelly CK. Global climate and the distribution of plant biomes. Philos Trans R Soc Lond B Biol Sci. 2004;359(1450):1465–1476. doi:10.1098/rstb.2004.1525

69. Oertel E. Beekeeping in the United States. In: USDA agriculture handbook. 335th ed. 1980. pp. 2–9.

70. Koo MH, Park YK. Investigation of flavonoid aglycones in propolis collected by two different varieties of bees in the same region. Biosci Biotechnol Biochem. 1997;61(2):367–369. doi:10.1271/bbb.61.367

71. Park YK, Alencar SM, Aguiar CL. Botanical origin and chemical composition of Brazilian propolis. J Agric Food Chem. 2002;50(9):2502–2506. doi:10.1021/jf011432b

72. Rubink WL, Luévano-Martínez P, Sugden A, Wilson WT, Collins AM. Subtropical Apis mellifera (Hymenoptera: Apidae) swarming dynamics and Africanization rates in Northeastern Mexico and Southern Texas. Ann Entomol Soc Am. 1996;89(2):243–251. doi:10.1093/aesa/89.2.243

73. García-Viguera C, Ferres F, Tomás-Barberán FA. Study of Canadian propolis by GC–MS and HPLC. Z Naturforsch. 1993;48c:731–735.

74. Christov R, Trusheva B, Popova M, Bankova V, Bertrand M. Chemical composition of propolis from Canada, its antiradical activity and plant origin. Nat Prod Res. 2006;20(6):531–536. doi:10.1080/14786410500056918

75. Bankova V, Galabov AS, Antonova D, Vilhelmova N, Di Perri B. Chemical composition of propolis extract ACF® and activity against herpes simplex virus. Phytomedicine. 2014;21(11):1432–1438. doi:10.1016/j.phymed.2014.04.026

76. Hoheisel O. The effects of Herstat (3% propolis ointment ACF) applicationin cold sores: a double-blind placebo-controlled clinical trial. J Clin Res. 2001;4:65–75.

77. Vynograd N, Vynograd I, Sosnowski Z. A comparative multi-centre study of the efficacy of propolis, acyclovir and placebo in the treatment of genital herpes (HSV). Phytomedicine. 2000;7(1):1–6. doi:10.1016/S0944-7113(00)80014-8

78. Cottica SM, Sabik H, Antoine C, Fortin J, Graveline N, Visentainer JV, Britten M. Characterization of Canadian propolis fractions obtained from two-step sequential extraction. LWT Food Sci Technol. 2012;60(1):609–614. doi:10.1016/j.lwt.2014.08.045

79. Lindenfelser LA. Antimicrobial activity of propolis. Am Bee J. 1967;107(3):90–92.

80. Lindenfelser LA. In vivo activity of propolis against Bacillus larvae. J Invertebr Pathol. 1968;12(1):129–131. doi:10.1016/0022-2011(68)90252-8

81. Johnson KS, Eischen FA, Giannasi DE. Chemical composition of North American bee propolis and biological activity towards larvae of greater wax moth (Lepidoptera: Pyralidae). J Chem Ecol. 1994;20(7):1783–1791. doi:10.1007/BF02059899.

82. Kumazawa S, Hamasaka T, Nakayama T. Antioxidant activity of propolis of various geographic origins. Food Chem. 2004;84(3):329–339. doi:10.1016/S0308-8146(03)00216-4

83. Sonmez S, Kirilmaz L, Yucesoy M, Yücel B, Yilmaz B. The effect of bee propolis on oral pathogens and human gingival fibroblasts. J Ethnopharmacol. 2005;102(3):371–376. doi:10.1016/j.jep.2005.06.035

84. Sawaya ACHF, Tomazela DM, Cunha IBS, Bankova V, Marcucci MC, Custodio AR, Eberlin MN. Electrospray ionization mass spectrometry fingerprinting of propolis. Analyst. 2004;129:739–744. doi:10.1039/B403873H

85. Aliboni A. Propolis from northern California and Oregon: chemical composition, botanical origin, and content of allergens. Z Naturforschung C. 2014;69(1–2):10–20. doi:10.5560/znc.2013-0114

86. Greenaway W, Scaysbrook T, Whately FR. The composition and plant origin of propolis: a report of work at Oxford. Bee World. 1990;71(3):107–118. doi:10.1080/0005772X.1990.11099047

87. Savka MA, Dailey L, Popova M, Mihaylova R, Merritt B, Masek M, Le P, Mat-Nor SR, Ahmad M, Hudson AO, Bankova V. Chemical composition and disruption of quorum sensing signaling in geographically diverse United States propolis. eCAM. 2015;2015:472593. doi:10.1155/2015/472593

88. Lotti C, Piccinelli AL, Arrevalo C, Ruiz I, de Castro GMM, De Sá LFR, Tessis AC, Ferreira-Pereira A, Rastrelli L. Constituents of Hondurian propolis with inhibitory effects on Saccharomyces cerevisiae multidrug resistance protein Pdr5p. J Agric Food Chem. 2012;60(42):10540–10545. doi:10.1021/jf302578r

89. Wollenweber E, Buchmann SL. Feral honey bees in the Sonoran Desert: propolis sources other than poplar (Populus spp.). Z Naturforsch. 1997;52c:530–535.

90. Inui S, Hosoya T, Kumazawa S. Hawaiian propolis: comparative analysis and botanical origin. Nat Prod Commun. 2014;9(2):165–166.

91. Brandon K, Gorenflo LJ, Rodrigues ASL, Waller RW. Reconciling biodiversity conservation, people, protected areas, and agricultural suitability in Mexico. World Dev. 2005;33(9):1403–1418. doi:10.1016/j.worlddev.2004.10.005

92. Li F, Yu-Min H, Awale S, Kadota S, Tezuka Y. Two new cytotoxic phenylallylflavanones from Mexican propolis. Chem Pharm Bull. 2011;59(9):1194–1196. doi:10.1248/cpb.59.1194

93. Navarro-Navarro M, Ruiz-Bustos P, Valencia D, Robles-Zepeda R, Ruiz-Bustos E, Virués C, Hernandez J, Domínguez Z, Velazquez C. Antibacterial activity of Sonoran propolis and some of its constituents against clinically significant Vibrio species. Foodborne Pathog Dis. 2013;10(2):150–158. doi:10.1089/fpd.2012.1318

94. Valencia D, Alday E, Robles-Zepeda R, Garibay-Escobar A, Galvez-Ruiz JC, Salas-Reyes M, Jiménez-Estrada M, Velazquez-Contreras E, Hernandez J, Velazquez C. Seasonal effect on chemical composition and biological activities of Sonoran propolis. Food Chem. 2012;131(2):645–651. doi:10.1016/j.foodchem.2011.08.086

95. Rivero-Cruz B, Martínez-Chávez A. Development and validation of a RP-HPLC method for the simultaneous quantification of flavonoids markers in Mexican propolis. Food Anal Methods. 2015;8(2):413–419. doi:10.1007/s12161-014-9908-5

96. Pino JA, Marbot R, Delgado A, Zumárraga C, Sauri E. Volatile constituents of propolis from honey bees and stingless bees from Yucatán. J Essent Oil Res. 2006;18(1):53–56. doi:10.1080/10412905.2006.9699384

97. Gebelein J. A geographic perspective of Cuban landscapes, Volume 15, Landscapes series. 1st ed. Miami, FL: Springer Netherlands, Springer Science+Business Media B.V.; 2012. 97 p. doi:10.1007/978-94-007-2406-8

98. Pascual C, Gonzalez R, Torricella RG. Scavenging action of propolis extract against oxygen radicals. J Ethnopharmacol. 1994;41(1–2):9–13. doi:10.1016/0378-8741(94)90052-3

99. González R, Remirez D, Rodriguez S, González A, Ancheta O, Merino N, Pascual, C. Hepatoprotective effects of propolis extract on paracetamol-induced liver damage in mice. Phytother Res. 1994;8(4):229–232.

100. Remirez D, González A, Rodríguez S, González R, Merino N, Ancheta O, Pascual C, Rojas E, Ramos EM. Efecto hepatoprotector del propoleo rojo

cubano en la toxicidad inducida por paracetamol en ratones. Rev Cuba Farm. 1994;28(1):45–54.

101. Remírez D, González R, Rodriguez S, Ancheta O, Bracho JC, Rosado A, Rojas E, Ramos ME. Protective effects of propolis extract on allyl alcohol-induced liver injury in mice. Phytomedicine. 1997;4(4):309–314.

102. González R, Corcho I, Remirez D, Ancheta O, Merino N, González A, Pascual C. Hepatoprotective effects of propolis extract on carbon tetrachloride-induced liver injury in rats. Phytother Res. 1995;9(2):114–117. doi:10.1002/ptr.2650090207

103. Merino N, González R, González A, Remírez D. Histopathological evaluation on the effect of red propolis on liver damage induced by CCl4 in rats. Arch Med Res. 1996;27(3):285–289.

104. Rodríguez S, Ancheta O, Ramos ME, Remirez D, Rojas E, Gonzalez R. Effects of Cuban red propolis on galactosamine-induced hepatitis in rats. Pharmacol Res. 1997;35(1):1–4.

105. Ledón N, Casaco A, González R, Merino N, González A, Tolón Z. Antipsoriatic, anti-inflammatory, and analgesic effects of an extract of red propolis. Acta Pharmacol Sin. 1997;18(3):474–276.

106. Cuesta-Rubio O, Cuellar A, Rojas N, Velez Castro H, Rastrelli L, Aquino R. A polyisoprenylated benzophenone from Cuban propolis. J Nat Prod. 1999;62(7):1013–1015. doi:10.1021/np980339n

107. Tomás-Barberán FA, Garcia-Viguera C, Vit-Olivier P, Ferreres F, Tomás-Lorente F. Phytochemical evidence for the botanical origin of tropical propolis from Venezuela. Phytochemistry. 1993;34(1):191–196. doi:10.1016/S0031-9422(00)90804-5

108. Cuesta-Rubio O, Velez-Castro H, Frontana-Uribe BA, Cárdenas J. Nemorosone, the major constituent of floral resins of Clusia rosea. Phytochemistry. 2001;57(2):279–283. doi:10.1016/S0031-9422(00)00510-0

109. Cuesta-Rubio O, Frontana-Uribe BA, Ramírez-Apan T, Cárdenas J. Polyisoprenylated benzophenones in Cuban propolis: biological activity of nemorosone. Z Naturforsch C. 2002;57(3–4):372–378.

110. Márquez Hernández I, Campo Fernández M, Cuesta-Rubio O, Piccinelli AL, Rastrelli L. Polyprenylated benzophenone derivatives from Cuban propolis. J Nat Prod. 2005;68(6):931–934. doi:10.1021/np0495884

111. Cuesta-Rubio O, Piccinelli AL, Campo Fernandez M, Márquez Hernández I, Rosad A Rastrelli L. Chemical characterization of Cuban propolis by HPLC–PDA, HPLC–MS, and NMR: the brown, red, and yellow Cuban varieties of propolis. J Agric Food Chem. 2007;55(18):7502–7509. doi:10.1021/jf071296w

112. Campo Fernández M, Cuesta-Rubio O, Rosado Pérez A, Montes de Oca Porto R, Márquez Hernández I, Piccinelli AL, Rastrelli L. GC–MS Determination of isoflavonoids in seven red Cuban propolis samples. J Agric Food Chem. 2008;56(21):9927–9932. doi:10.1021/jf801870f

113. Márquez Hernández I, Cuesta-Rubio O, Campo Fernández M, Rosado Pérez A, Montes de Oca Porto R, Piccinelli AL, Rastrelli L. Studies on the constituents of yellow Cuban propolis: GC–MS determination of triterpenoids and flavonoids. J Agric Food Chem. 2010;58(8):4725–4730. doi:10.1021/jf904527n

114. Díaz-Carballo D, Malak S, Bardenheuer W, Freistuehler M, Reusch HP. Cytotoxic activity of nemorosone in neuroblastoma cells. J Cell Mol Med. 2008;12(6B):2598–2608. doi:10.1111/j.1582-4934.2008.00232.x

115. Popolo A, Piccinelli LA, Morello S, Cuesta-Rubio O, Sorrentino R, Rastrelli L, Pinto A. Antiproliferative activity of brown Cuban propolis extract on human breast cancer cells. Nat Prod Commun. 2009;4(12):1711–1716.

116. Popolo A, Piccinelli AL, Morello S, Sorrentino R, Osmany CR, Rastrelli L, Pinto A. Cytotoxic activity of nemorosone in human MCF-7 breast cancer cells. Can J Physiol Pharmacol. 2011;89(1):50–57. doi:10.1139/y10-100

117. Camargo MS, Prieto AM, Resende FA; Boldrin PK, Cardoso CR, Fernández MF, Molina-Molina JM, Olea N, Vilegas W, Cuesta-Rubio O, Varanda EA. Evaluation of estrogenic, antiestrogenic and genotoxic activity of nemorosone, the major compound found in brown Cuban propolis. BMC Complement Altern Med. 2013;13:201. doi:10.1186/1472-6882-13-201

118. Díaz-Carballo D, Malak S, Bardenheuer W, Freistuehler M, Peter Reusch H. The contribution of plukenetione A to the anti-tumoral activity of Cuban propolis. Bioorg Med Chem. 2008;16(22):9635–9643. doi:10.1016/j.bmc.2008.10.019

119. Diaz-Carballo D, Freistã Hler M, Malak S, Bardenheuer W, Reusch HP. Mucronulatol from Caribbean propolis exerts cytotoxic effects on human tumor cell lines. Int J Clin Pharmacol Ther. 2008;46(5):226–235.

120. Camargo MS, Varela SD, de Oliveira AN, Resende FA, Cuesta-Rubio O, Vilegas W, Varanda EA. Assessment of the estrogenic, mutagenic and antimutagenic activity of nemorosone. Rev Bras Farmacogn. 2011;21(5):921–927. doi:10.1590/S0102-695X2011005000148

121. Daugsch A, Moraes CS, Fort P, Park YK. Brazilian red propolis—chemical composition and botanical origin. eCAM. 2008;5(4):435–441. doi:10.1093/ecam/nem057

122. Silva BB, Rosalen PL, Cury JA, Ikegaki M, Souza VC, Esteves A, Alencar SM. Chemical composition and botanical origin of red propolis, a new type of Brazilian propolis. eCAM. 2008;5(3):313–331. doi:10.1093/ecam/nem059

123. Awale S, Li F, Onozuka H, Esumi H, Tezuka Y, Kadota S. Constituents of Brazilian red propolis and their preferential cytotoxic activity against human pancreatic PANC-1 cancer cell line in nutrient-deprived condition. Bioorg Med Chem. 2008;16(1):181–189. doi:10.1016/j.bmc.2007.10.004

124. Trusheva B, Popova M, Bankova V, Simova S, Marcucci M, Miorin P, Rocha Pasin F, Tsvetkova I. Bioactive constituents of Brazilian red propolis. eCAM. 2006;3(2):249–254. doi:10.1093/ecam/nel006

125. Piccinelli AL, Lotti C, Campone L, Cuesta-Rubio O, Campo Fernandez M, Rastrelli. Cuban and Brazilian red propolis: botanical origin and comparative analysis by high-performance liquid chromatography-photodiode array detection/electrospray ionization tandem mass spectrometry. J Agric Food Chem. 2011;59(12):6484–6491. doi:10.1021/jf201280z

126. Popova M, Bankova V, Spassov S, Tsvetkova I, Naydenski C, Vides Silva M, Tsartsarova, M. New bioactive chalcones in propolis from El Salvador. Z Naturforsch C. 2001;56(7–8):593–596.

127. Popova M, Bankova V, Tsvetkova I, Naydenski C, Silva MV. The first glycosides isolated from propolis: diterpene rhamnosides. Z Naturforsch C. 2001;56(11–12):1108–1111.

128. Wilf P, Cúneo NR, Johnson KR, Hicks JF, Wing SL, Obradovich JD. High plant diversity in Eocene South America: evidence from Patagonia. Science. 2003;300(5616):122–125.

129. Young KR, Ulloa-Ulloa C, Luteyn JL, Knapp S. Plant evolution and endemism in Andean South America: an introduction. Bot Rev. 2002;68(1):4–21.

130. Marchant R, Behling H, Berrio JC, Hooghiemstra H, van Geel B, van der Hammen T, Herrera L, Melief B, van Reenen G, Wille M. Vegetation disturbance and human population in Colombia—a regional reconstruction. Antiquity. 2004;78(302):828–838. doi:10.1017/S0003598X0011347X

131. Meneses EA, Durango DL, García CM. Antifungal activity against postharvest fungi by extracts from Colombian propolis. Quím Nova. 2009;32(8):2011–2017. doi:10.1590/S0100-40422009000800006

132. Piccinelli AL, Mencherini T, Celano R, Mouhoubi Z, Tamendjari A, Aquino RP, Rastrelli L. Chemical composition and antioxidant activity of Algerian propolis. J Agric Food Chem. 2013;61(21):5080–5088. doi:10.1021/jf400779w

133. Palomino García LR, Martínez Galán JP, García Pajón CM, Gil González JH, Durango Restrepo DL. Physicochemical characterization and antimicrobial activity of propolis from municipality of La Union (Antioquia, Colombia). Rev Fac Nal Agr Medellín. 2010;63(1):5373–5383.

134. Trusheva B, Popova M, Naydenski H, Tsvetkova I, Rodriguez JG, Bankova V. New polyisoprenylated benzophenones from Venezuelan propolis. Fitoterapia. 2004;75(7–8):683–689.

135. Trusheva B, Popova M, Kortenska V, Totseva I, Rodriguez JG, Bankova V. Chemical composition and antioxidant activity of propolis from Venezuela. Oxid Commun. 2005;28(2):490–496.

136. Banskota AH, Tezuka Y, Adnyana IK, Midorikawa K, Matsushige K, Message D, Huertas AAG; Kadora S. Cytotoxic, hepatoprotective and free radical scavenging effects of propolis from Brazil, Peru, the Netherlands and China. J Ethnopharmacol. 2000;72(1–2):239–246. doi:10.1016/S0378-8741(00)00252-X

137. Nina N, Quispe C, Jiménez-Aspee F, Theoduloz C, Giménez A, Schmeda-Hirschmann G. Chemical profiling and antioxidant activity of Bolivian propolis. J Sci Food Agric. 2016;96(6):2142–2153. doi:10.1002/jsfa.7330

138. Chang R, Piló-Veloso D, Morais SAL, Nascimento EA. Analysis of a Brazilian green propolis from Baccharis dracunculifolia by HLPC–APCI–MS and GS–MS. Braz J Pharmacog Rev. 2008;18(4):549–556. doi:10.1590/S0102-695X2008000400009

139. Nina N, Quispe C, Jiménez-Aspee F, Theoduloz C, Giménez A, Schmeda-Hirschmann G. Antibacterial and leishmanicidal activity of Bolivian propolis. Lett Appl Microbiol. 2016;62(3):290–296. doi:10.1111/lam.12543

140. Giulietti AM, Harley RM, De Queiroz LP, Wanderley MDGL, Van der Berg C. Biodiversity and conservation of plants in Brazil. Conserv Biol. 2005;19(3):632–639. doi:10.1111/j.1523-1739.2005.00704.x

141. Aga H, Shibuya T, Sugimoto T, Kurimoto M, Nakajima S. Isolation and identification of antimicrobial compounds in Brazilian propolis. Biosci Biotechnol Biochem. 1994;58(5):945–946. doi:10.1271/bbb.58.945

142. Tazawa S, Warashima T, Noro T, Miyase T. Studies on the constituents of Brazilian propolis. Chem Pharm Bull. 1998;46(9):1477–1479. doi:10.1248/cpb.46.1477

143. Tazawa S, Warashina T, Noro T. Studies on the constituents of Brazilian propolis. II. Chem Pharm Bull. 1999;47(10):1388–1392. doi:10.1248/cpb.47.1388

144. Banskota AH, Tezuka Y, Midorikawa K, Matsushige K, Kadota S. Two novel cytotoxic benzofuran derivatives from Brazilian propolis. J Nat Prod. 2000;63:1277–1279. doi:10.1021/np000143z

145. Park YK, Ikegaki M, Alencar SM, Moura FF. Evaluation of Brazilian propolis by both physicochemical methods and biological activity. Honeybee Sci. 2000;21:85–90.

146. Park YK, Ikegaki M, Alencar SM. Classification of Brazilian propolis by physicochemical method and biological activity. Mensagem Doce. 2000;58:2–7.

147. Park YK, Alencar SM, Scamparini ARP, Aguiar CL. Propolis produced in Southern Brazil, Argentine and Uruguay: phytochemical evidence for the plant origin. Cienc Rural, Santa Maria. 2002;35:997–1003.

148. Castro ML, do Nascimento AM, Ikegaki M, Costa-Neto CM, Alencar SM, Rosalen PL. Identification of a bioactive compound isolated from Brazilian propolis type 6. Bioorg Med Chem. 2009;17(14):5332–5335. doi:10.1016/j.bmc.2009.04.066

149. Alencar SM, Oldoni TLC, Castro ML, Cabral ISR, Costa-Neto CM, Cury JA, Rosalen PL, Ikegaki M. Chemical composition and biological activity of a new type of Brazilian propolis: red propolis. J Ethnopharmacol. 2007;113(2):278–283. doi:10.1016/j.jep.2007.06.005

150. Oliveira AP, França HS, Kuster RM, Teixeira LA, Rocha LM. Chemical composition and antibacterial activity of Brazilian propolis essential oil. J Venom Anim Toxins Incl Trop Dis. 2010;16(1):121–130. doi:10.1590/S1678-91992010005000007

151. Murad JM, Calvi SA, Soares AMVC, Bankova V, Sforcin JM. Effects of propolis from Brazil and Bulgaria on fungicidal activity of macrophages against Paracoccidioides brasiliensis. J Ethnopharmacol. 2002;79(3):331–334. doi:10.1016/S0378-8741(01)00404-4

152. Sforcin JM. Propolis and the immune system: a review. J Ethnopharmacol. 2007;113(1):1–14. doi:10.1016/j.jep.2007.05.012

153. Orsi RO, Funari SRC, Soares AMVC, Calvi SA, Oliveira SL, Sforcin JM, Bankova V. Immunomodulatory action of propolis on macrophage activation. J Venom Anim Toxins. 2000;6(2):205–219. doi:10.1590/S0104-79302000000200006

154. Orsatti CL, Missima F, Pagliarone AC, Bachiega TF, Búfalo MC, Araújo JP, Sforcin JM. Propolis immunomodulatory action in vivo on Toll-like receptors 2 and 4 expression and on pro-inflammatory cytokines production in mice. Phytother Res. 2010;24(8):1141–1146. doi:1141-6. doi:10.1002/ptr.3086

155. Sforcin JM, Orisi RO, Bankova V. Effect of propolis, some isolated compounds and its source plant on antibody production. J Ethnopharmacol. 2005;98(3):301–305. doi:10.1016/j.jep.2005.01.042

156. Pagliarone AC, Missima F, Orsatti CL, Bachiega TF, Sforcin JM. Propolis effect on Th1/Th2 cytokines production by acutely stressed mice. J Ethnopharmacol. 2009;125(2):230–233. doi:10.1016/j.jep.2009.07.005

157. Pagliarone AC, Orsatti CL, Búfalo MC, Missima F, Bachiega TF, Araújo JP, Sforcin JM. Propolis effects on pro-inflammatory cytokine production and Toll-like receptor 2 and 4 expression in stressed mice. Int Immunopharmacol. 2009;9(11):1352–1356. doi:10.1016/j.intimp.2009.08.005

158. Missima F, Sforcin JM. Green Brazilian propolis action on macrophages and lymphoid organs of chronically stressed mice. eCAM. 2008;5(1):71–75. doi:10.1093/ecam/nel112

159. Missima F, Pagliarone AC, Orsatti CL, Sforcin JM. The effect of propolis on pro-inflammatory cytokines produced by melanoma-bearing mice submitted to chronic stress. J ApiProduct ApiMedical Sci. 2009;1(1):11–15. doi:10.3896/IBRA.4.01.1.05

160. Missima F, Pagliarone AC, Orsatti CL, Araújo JP, Sforcin JM. Propolis effect on Th1/Th2 cytokines' expression and production by melanoma-bearing mice submitted to stress. Phytother Res. 2010;24(10):1501–1507. doi:10.1002/ptr.3142

161. Búfalo MC, Bordon-Graciani AP, Conti BJ, de Assis GM, Sforcin JM. The immunomodulatory effect of propolis on receptors expression, cytokine production and fungicidal activity of human monocytes. J Pharm Pharmacol. 2014;66(10):1497–1504. doi:10.1111/jphp.12279

162. Búfalo MC, Sforcin JM. The modulatory effects of caffeic acid on human monocytes and its involvement in propolis action. J Pharm Pharmacol. 2015;67(5):740–745. doi:10.1111/jphp.12364

163. Conti BJ, Búfalo MC, de Assis GM, Bankova V, Sforcin JM. Cinnamic acid is partially involved in propolis immunomodulatory action on human monocytes. eCAM. 2013;2013:109864. doi:10.1155/2013/109864

164. Búfalo MC, Ferreira I, Costa G, Francisco V, Liveral J, Cruz MT, Lopes MC, Batista MT, Sforcin JM. Propolis and its constituent caffeic acid suppress LPS-stimulated pro-inflammatory response by blocking NF-κB and MAPK activation in macrophages. J Ethnopharmacol. 2013;149(1):84–92. doi:10.1016/j.jep.2013.06.004

165. Búfalo MC, Candeias JMG, Sousa JPB, Bastos JK, Sforcin JM. In vitro cytotoxic activity of Baccharis dracunculifolia and propolis against HEp-2 cells. Nat Prod Res. 2010;24(18):1710–1718. doi:10.1080/14786410902909300

166. Búfalo MC, Candeias JMG, Sforcin JM. In vitro cytotoxic effect of Brazilian green propolis on human laryngeal epidermoid carcinoma (HEp-2) cells. eCAM. 2009;6(4):483–487. doi:10.1093/ecam/nem147

167. Cinegaglia NC, Bersano PRO, Búfalo MC, Sforcin JM. Cytotoxic action of Brazilian propolis in vitro on canine osteosarcoma cells. Phytother Res. 2012;27(9):1277–1281. doi:10.1002/ptr.4861

168. Ishihara M, Naoi K, Hashita M, Itoh Y, Suzui M. Growth inhibitory activity of ethanol extracts of Chinese and Brazilian propolis in four human colon carcinoma cell lines. Oncol Rep. 2009;22(2):349–354.

169. Rodríguez C, Leoni E, Lezama F, Altesor A. Temporal trends in species composition and plant traits in natural grasslands of Uruguay. J Veg Sci. 2003;14(3):433–440. doi:10.1111/j.1654-1103.2003.tb02169.x

170. Bonvehí JS, Coll FV. Study on propolis quality from China and Uruguay. Z Naturforsch C. 2000;55(9–10):778–784. doi:10.1515/znc-2000-9-1017

171. Kumazawa S, Hayashi K, Kajiya K, Ishii T, Hamasaka T, Nakayama T. Studies of the constituents of Uruguayan propolis. J Agric Food Chem. 2002;50(17):4777–4782. doi:10.1021/jf020279y

172. Silva V, Genta G, Möller MN, Masner M, Thmson L, Romero N, Radi R, Fernandes DC, Laurindo FRM, Heinzen H, Fierro W, Denicola A. Antioxidant activity of Uruguayan propolis. In vitro and cellular assays. J Agric Food Chem. 2011;59(12):6430–6437. doi:10.1021/jf201032y

173. Kaškonienė V, Kaškonas P, Maruška A, Kubilienė L. Chemometric analysis of volatiles of propolis from different regions using static headspace GC–MS. Cent Eur J Chem. 2014;12(6):736–746. doi:10.2478/s11532-014-0521-7

174. Bucher EH. Revista Chilena de Historia Natural. Herbivory in arid and semi-arid regions of Argentina. Rev Chil Hist Nat. 1987;60:265–273.

175. Fernández OA, Busso CA. Arid and Semi-arid Rangelands: Two Thirds of Argentina. In: Arnalds O, Archer S, editors. Case studies of rangeland desertification. Agricultural Research Institute Report No. 200, Reykjavik. 1999. pp. 41–60.

176. Isla MI, Paredes-Guzman JF, Nieva-Moreno MI, Koo H, Park YK. Some chemical composition and biological activity of Northern Argentine propolis. J Agric Food Chem. 2005;53(4):1166–1172. doi:10.1021/jf040130h

177. Nieva-Moreno MI, Zampini IC, Ordóñez RM, Jaime GS, Vattuone MA, Isla MI. Evaluation of the Cytotoxicity, Genotoxicity, Mutagenicity, and Antimutagenicity of Propolis from Tucuman, Argentina. J. Agric. Food Chem. 2005;53:8957–8962. doi:10.1021/jf0513359

178. Quiroga EN, Sampietro DA, Sobcrón JR, Sgariglia MA, Vattuone MA. Propolis from the northwest of Argentina as a source of antifungal principles. J Appl Microbiol. 2006;101:103–110. doi:10.1111/j.1365-2672.2006.02904.x

179. Agüero MB, Gonzalez M, Lima B, Svetaz L, Sánchez M, Zacchino S, Feresin GE, Schmeda-Hirschmann G, Palermo J, Wunderlin D, Tapia A. Argentinean propolis from Zuccagnia punctata Cav. (Caesalpinieae) exudates:

phytochemical characterization and antifungal activity. J Agric Food Chem. 2010;58(1):194–201. doi:10.1021/jf902991t.

180. Vera N, Solorzano E, Ordoñez R, Maldonado L, Bedascarrasbure E, Isla MI. Chemical composition of Argentinean propolis collected in extreme regions and its relation with antimicrobial and antioxidant activities. Nat Prod Commun. 2011;6(6):823–827.

181. Kumazawa S, Ahn M, Fujimoto T, Kato M. Radical-scavenging activity and phenolic constituents of propolis from different regions of Argentina. Nat Prod Res. 2010;24(9):804–812. doi:10.1080/14786410802615270

182. Solórzano E, Vera N, Cuello S, Ordoñez R, Zampini C, Maldonado L, Bedascarrasbure E, Isla MI. Chalcones in bioactive Argentine propolis collected in arid environments. Nat Prod Commun. 2012;7(7):879–882.

183. Agüero MB, Svetaz L, Sánchez M, Luna L, Lima B, López ML, Zacchino S, Palermo J, Wuderlin D, Feresin GE, Tapia A. Argentinean Andean propolis associated with the medicinal plant Larrea nitida Cav. (Zygophyllaceae). HPLC–MS and GC–MS characterization and antifungal activity. Food Chem Toxicol. 2011;49(9):1970–1978. doi:10.1016/j.fct.2011.05.008

184. Agüero MA, Svetaz L, Baroni V, Lima B, Luna L, Zacchino S, Saavedra P, Wunderlin D, Feresin GE, Tapia A. Urban propolis from San Juan province (Argentina): ethnopharmacological uses and antifungal activity against Candida and dermatophytes. Ind Crops Prod. 2014;57:166–173. doi:10.1016/j.indcrop.2014.03.009

185. Gardana C, Scaglianti M, Pieta P, Simonetti P. Analysis of the polyphenolic fraction of propolis from different sources by liquid chromatography–tandem mass spectrometry. J Pharm Biomed Anal. 2007;45(3):390–399. doi:10.1016/j.jpba.2007.06.022

186. Volpi N, Bergonzini G. Analysis of flavonoids of propolis by on-line HPLC-electrospray mass spectrometry. J Phram Biomed Anal. 2006;42:354–361. doi:10.1016/j.jpba.2006.04.017

187. Chaillou LL, Nazareno MA. Chemical variability in propolis from Santiago del Estero, Argentina, related to the arboreal environment as the sources of resins. J Sci Food Agric. 2009;89(6):978–983. doi:10.1002/jsfa.3542

188. Chaillou LL, Nazareno MA. Bioactivity of propolis from Santiago del Estero, Argentina, related to their chemical composition. LWT Food Sci Technol. 2009;49(8):1422–1427. doi:10.1016/j.lwt.2009.03.002

189. Castro SA, Figueroa JA, Muñoz-Shick M, Jaksic FM. Minimum residence time, biogeographical origin, and life cycle as determinants of the

geographical extent of naturalized plants in continental Chile. Divers Distrib. 2005;11(3):183–191. doi:10.1111/j.1366-9516.2005.00145.x

190. ValcicS, Montenegro G, Timmermann BN. Lignans from Chilean propolis. J Nat Prod. 1998;61(6):771–775. doi:10.1021/np980017j

191. Muñoz O, Peña RC, Ureta E, Montenegro G, Timmermann BN. Propolis from Chilean Matorral Hives. Z Naturforsch C. 2001;56(3–4):269–272.

192. Muñoz O, Peña RC, Ureta E, Montenegro GM, Cladwell C, Timmermann BN. Phenolic compounds of propolis from Central Chilean Matorral. Z Natusforsch C. 2001;56(3–4):273–277.

193. Castro C, Mura F, Valenzuela G, Figueroa C, Salinas R, Zuñiga MC, Torres JL, Fuguet E, Delporte C. Identification of phenolic compounds by HPLC-ESI-MS/MS and antioxidant activity from Chilean propolis. Food Res Intern. 2014;64:873–879. doi:10.1016/j.foodres.2014.08.050

194. Valenzuela-Barra G, Castro C, Figueroa C, Barriga A, Silva X, de las Heras B, Hortelano S, Delporte C. Anti-inflammatory activity and phenolic profile of propolis from two locations in Región Metropolitana de Santiago, Chile. J Ethnopharmacol. 2015;168:37–44. doi:10.1016/j.jep.2015.03.050

195. Russo A, Cardile V, Sanchez F, Troncoso N, Vanella A, Garbarino JA. Chilean propolis: antioxidant activity and antiproliferative action in human tumor cell lines. Life Sci. 2004;76(5):545–558.

196. Barrientos L, Herrera CL, Montenegro G, Ortega X, Veloz J, Alvear M, Cuevas A, Saavedra N, Salazar L. Chemical and botanical characterization of Chilean propolis and biological activity on cariogenic bacteria Streptococcus mutans and Streptococcus sobrinus. Braz J Microbiol. 2013;44(2):577–585.

197. Veloz JJ, Saavedra N, Lillo A, Alvear M, Barrientos L, Salazar LA. Antibiofilm activity of Chilean propolis on Streptococcus mutans is influenced by the year of collection. BioMed Res Int. 2015;2015:291351.

06 ADVANCEMENTS IN FRUIT TREE POLLINATION TECHNOLOGY AND INDUSTRIALIZATION IN CHINA

1. INTRODUCTION:AN OVERVIEW

Approximately 75% of all crops rely on bees for pollination. Pollination of certain crops is facilitated by domesticated honeybees, whereas the majority of agricultural pollination is carried out by wild insects, such as wild bees. Wild bee diversity in China has decreased, leading farmers to rely on controlled bees like Apis mellifera. In southwest China, there is a significant example of the expansion of cultivated areas for apple and pear trees. However, this expansion has resulted in a drop in the population of wild pollinators due to environmental deterioration, air pollution, pesticide usage, and other factors. Consequently, crops are unable to receive adequate pollination. Regions experiencing a significant lack of pollinators resort to employing human labor as a substitute for insect pollination, resulting in a substantial rise in production costs. This study aims to minimize the expenses associated with pollination and enhance its effectiveness by investigating strategies to optimize bee pollination of apple and pear trees. It explores various elements, such as methods to attract bees for pollination, the selection of the most suitable bee species, and the ideal arrangement of bees for optimal pollination. This work establishes a foundation for the utilization of bee pollination technology in agricultural settings.

2. POLLINATION PROBLEM FACING DEVELOPMENT OF THE FRUIT INDUSTRY IN CHINA

2.1. Development of the fruit industry

China is a vast nation that boasts numerous diverse ecosystems and abundant fruit tree reserves. Fruit tree agriculture encompasses about 50 families, with over 300 species and exceeding 10,000 variations. The main fruits and nuts widely distributed throughout China include apple, pear, peach, plum, apricot, grapes, cherry, walnuts, citrus fruit, lychee, longan, loquat, olives, kiwi fruit, fig, pomegranate,

Phyllanthus emblica (Indian gooseberry), banana, pineapple, durian, mangosteen, breadfruit, cocoa, and betel nuts. The region dedicated to fruit cultivation and fruit production is among the highest globally. Apple trees exhibit a broad geographical range in China, displaying both evergreen and deciduous characteristics, and are extensively cultivated over the Yunnan-Guizhou plateau. The deciduous varieties can be broken down into temperate deciduous, deciduous with dry temperature, dry cold deciduous, and hardy deciduous (figure 1)[1]. In 2011 the cultivated area devoted to apples in China reached 2.177 million hm^2, accounted for about 42.0 of the world's total area devoted to the cultivation of apple, total output reached 35.985 million t, represented about 54.2 of total world yield, and had an output value of 160.52 billion yuan[2].

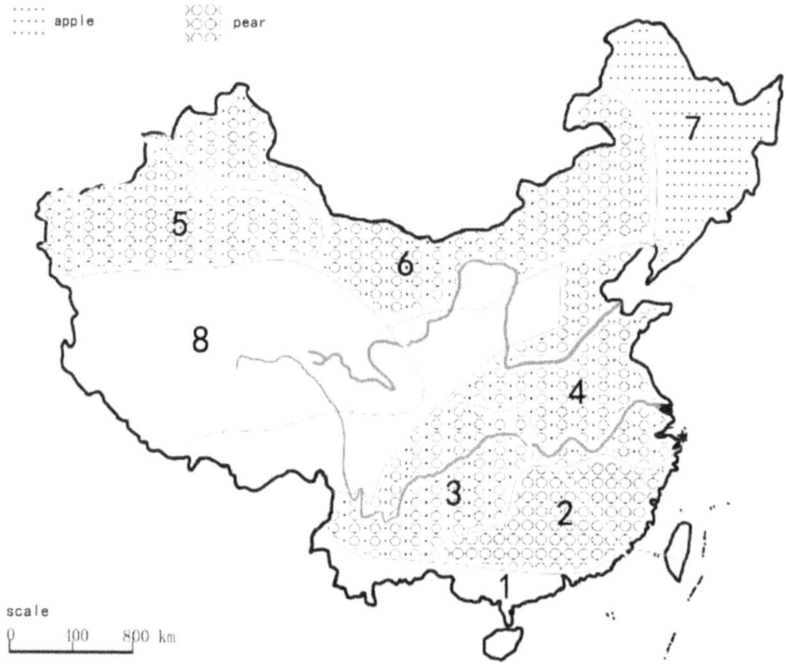

Figure 1. Distribution of fruit trees in China.

Pear trees are also important; they are grown across five fruit zones. They too can be deciduous or evergreen. Subtropical evergreen trees are grown in the south, whereas cold and dry deciduous fruit trees are grown in the north[1] (figure 1). By the end of 2012, according to statistical data released by the United Nations Food and Agriculture Organization[3], China's harvest pear covers an area of 1.138 million hm^2, production is 1.721 million t, and area and yield are among the highest in the world.

Peach trees grow in temperate areas of China. They are mainly distributed in the YunnanGuizhou Plateau. They too can be evergreen or deciduous. They can be broken down into evergreen and deciduous mixed, temperate deciduous, drytemperature deciduous (figure 1)[1]. According to statistical data published by the United Nations Food and Agriculture Organization[3] by the end of 2010 the area devoted to peach cultivation in China was 732,000 hm^2 and yield reached 10.828 million t, both of which were the highest in the world.

In recent years, China has experienced significant advancements in fruit production. This mostly pertains to the production of larger cultivars, the process of breeding, the enhancement of fruit quality, and the promotion of well-liked market variants. The biological properties of tree species, their capacity to adapt to the environment, the enhancement of plant types, and the cultivation of trees in the most suitable biome all contribute to maximizing efficiency[4, 5] and enhancing the competitiveness of the fruit market, leading to economic gains.

There is a significant scarcity of pollinators for fruit trees.

Most of the apple, pear, plum, apricot, and chestnut varieties as well as almost all the sweet cherry varieties need be crosspollinated for the production of seed[6]. Because the pollen grains of nuts and fruits are big, heavy, sticky, and have an outer wall with a pattern of bumps, the wind cannot spread them easily. These fruits rely mainly on insect pollination, especially bees[7 8].

According to a survey of the literature, in the major appleproducing areas in Shanxi Province a total of 23 species of insects visit apple flowers. They belong to 4 orders and 14 families, mainly comprising hymenopteran (Hymenoptera) bees (Apidae), an anthophorid bee (Anthophoridae), leafcutting bees (Megachilidae), an andrenid bee (Andrenidae), and a dipteran (Diptera, Syrphidae). Among them, the Italian bee (*Apis mellifera ligustica*), which accounts for 61.5 to 99.4 of pollinators, is the apple's main pollinator. The foraging peaks of the Italian bee and *Anthophora plumipes* (Pallas, 1772), another local pollinator known as the hairyfooted flower bee, stagger, reducing competition between each other[9]. Lu Yanguo et al. visited insectpollinated apple blossom in central and southern loess plateau regions. The results show that the bee is the main pollinator in Tianshui and Liangdi, where it accounted for 92.6 and 60, respectively, of all insect pollination[10]. Yang and Wu surveyed the number and species of insects pollinating kiwi fruit flowers. They identified 16 species of pollinator, including bees (11), food aphid flies (4), and a dung beetle (1). Statistical analysis of the pollination behavior and pollination frequency showed that the bee *Apis cerana cerana* and the Italian bee are the best pollinators, with other insects much less active, carrying less pollen, and having much less of an impact[11]. Zhang Yunyi et al. investigated species of pollinators and the quantity of large cherry trees in Shanxi, and found that hymenopterans accounted for 64.83 to 74.81 of the pollinators. Hymenopteran pollinators are the most important. Interestingly, pollinators in mountainous area are richer in species than those in the plain[12].

There has been a significant decline in the population of bees and other pollinators in recent years, attracting global attention. The probable cause for the decrease in the population of wild pollinators is the utilization of pesticides[13, 14]. Furthermore, the extensive deforestation for agricultural purposes throughout the 15th and 16th centuries resulted in significant limitations on the natural habitat of wild insects. Climate change can lead to variations in plant phenophases and pollinator development cycles, which can result in insufficient pollination[17]. To summarize, the main issue in fruit tree pollination is the urgent need for artificial pollination in certain regions, as shown in figure 2. For instance, around 30% of China's pear trees rely on artificial pollination[18].

Figure 2. Artificial pollination.

3. THE CURRENT STATUS OF POLLINATION TECHNOLOGY IN THE FRUIT SECTOR IN CHINA

3.1. Chinese institutions engaged in research on fruit tree pollination technology

Between 1980 and 2013, Chinese literature shows that there are 161 institutions involved in researching the pollination of fruit trees. The most important ones include the Chinese Academy of Agricultural Sciences Institute of Bees, the Horticulture Institute of the Shanxi Academy of Agricultural Sciences, and the Beijing Academy of Agriculture and Forestry. From 1980 to 2011, a total of 324 research papers were published on the topic of bees pollinating trees (figure 3). Among these, 66 articles were written between 1980 and 1992, with an average of 5 articles per year. The number of articles remained relatively consistent between 1993 and 2001. However, there was a significant increase in the number of publications from 2001 to 2011, with a total of 204 articles published during that period. The research topic encompassed various aspects, including bee pollination methodologies, bee species selection, and the impact on crops and pollination assessment. Several studies examined the impact of bee pollination on production, the relationship between bee pollination and ecology, crop breeding, and the management of pollination colonies (see figure 4).

Figure 3. Literature about bee pollination from 1980 to 2011.

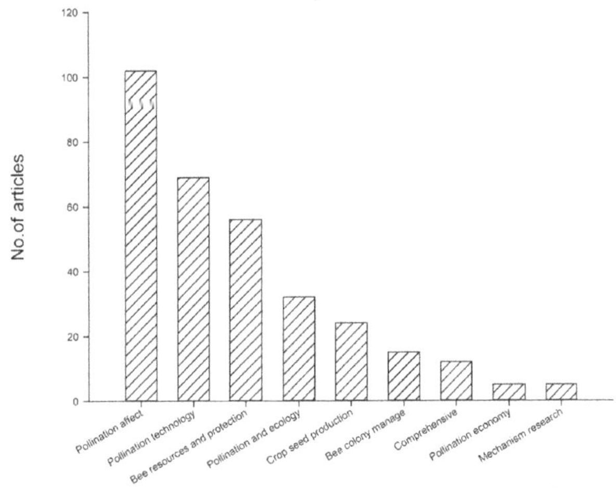

Figure 4. Bee pollination literature by research content.

Between 2001 and 2011, nine books were published on insect pollination technology research and application (table 1).

Table 1. Monographs on bee pollination in China.

Monograph title	Author	Publishing date	Press	Page extent
Technology of Bee Breeding and Pollination	Zeng Zhijiang	2001.07	Shanghai Popular Science Press	274
Bee Pollination	Shao Youquan	2001.09	Shanxi Science and Technology Press	113
New Technology of Osmia Pollination for Fruit Tree	Zhou Weiru	2002.01	Jindun Press	143

Monograph title	Author	Publishing date	Press	Page extent
Pollination Insect and Technology of Pollination	Wu Jie	2004.01	Chinese Agric ulture Press	204
Bee Pollination Handbook	Zhang Zhongyin, An Jiandong, Luo Shudong et al.	2008.10	Chinese Agri culture Press	93
Apiculture and Ecology	Wang Yong	2009.05	Chinese Agriculture Science and Technology Press	127
Yield Increasing Techniques of Insect Pollination for Fruit and Vegetable	Shao Youquan, Qi Haiping	2010.05	Jindun Press	202
Bumblebee Artificial Propagation and Pollination	Liu Xinyu, Gao Chongdong	2011.05	Northwest Agric ulture and Forestry University Press	133
Technique of Crop Yield Increasing: Bee Pollination	Wu Jie, Shao Youquan	2011.07	Chinese agriculture press	173

According to the patent database of the State Intellectual Property Office, there were a total of 54 categories of patents related to bee pollination between 2000 and 2013. Out of them, 43 were invention patents and 11 were utility model patents. The topics discussed included various aspects related to bee pollination, such as application technologies, breeding techniques, management methods, induction methods, hive design and transformation, and control devices for bee pollination. The patentees originated from various regions and cities, with Beijing, Zhejiang, and Shandong being the top three in terms of numbers. From 2000 to 2007, a cumulative of 15 patents were awarded, whereas from 2008 to 2013, the cumulative number increased to 39. There is a significant upward trend in patents relating to bee pollination.

We obtained 41 categories of bee pollination from the National Network of Scientific and Technological Achievements. The most significant among them were 17 categories focused on the breeding and selection of bee species, which accounted for 41.5% of the overall findings.

In addition, there were classes covering bee application and technology research (14), beehive design (2), and pollinating bee species resources (3). The results show that of the pollinating beesbumblebees, osmia bees, and leafcuttersbumblebee research was the most impotant. From the point of view of achievements, institutes in Beijing, Jilin, and Shanxi were in the top three. Regarding the number of achievements the Beijing Forestry Academy of Sciences, the Chinese Academy of Agricultural Sciences Institute of Bees, and the Horticulture Institute of the Shanxi Academy of Agricultural Sciences were in the top three. The Beijing Academy of Agriculture and Forestry has made significant advancements in the establishment of facilities for agricultural pollination, bee species breeding, usage, demonstration, and the development of pollination hives. The Chinese Academy of Agricultural Sciences Institute of Bees achieved a significant advancement in the breeding, exploitation, and application of bumblebees. The Horticulture Institution of the Shanxi Academy of Agricultural Sciences achieved significant achievement by enhancing its facilities for study and application of vegetable producing technologies.

3.2. Effectiveness evaluation of bee pollination for fruit trees

The introduction of the bee species Osmia cornifrons (Ra doszkowski) by the Biological Control Research Institute of the Chinese Academy of Agricultural Sciences from Japan in 1987 had a significant positive impact on the pollination effectiveness, fruit-setting rate, and fruit quality of apricot, cherry, peach, pear, and apple in the provinces of Hebei and Shandong.

After employing Apis mellifera ligustica for the purpose of apple pollination, Zhang Guiqian et al. observed that bee pollination resulted in a 46.8% increase in Red Fuji apple fruit yield, a 22.4% decrease in the rate of deformed fruit, and a yield of 14,124 kg/hm^2 [20], surpassing the results of natural pollination. He Weizhi and Zhou Weiru researched the use of the concavelipped bee *Osmia excavata* Alfken, the Italian bee, and artificial pollination for Red Fuji apple pollination. The results showed the apple fruit yield of the six kinds of pollination was significantly higher than that of natural pollination; the pollination effect of osmia bees combined with Italian bees was best with a high inflorescence fruit rate of 99.6 [21]. Lou Delong et al. found that the Red Fuji apple fruit yield, production, and coloring index of bee pollination were higher than those of natural pollination by 15, 36.26, and 17.07, respectively[22]. Using bee pollination for apple and pear, Zhao Zhonghua et al. found that fruit yield was more than 20 higher than artificial pollination and the average production of each acre was 224.4 kg, 335.3 kg with the increase rate of 8.7 and 11.3 [23]. Yuan Feng et al. used osmia bees and honeybees to pollinate Red Fuji apple trees and found that fruit yield increased by 14.68 and 10.95, respectively, over the natural pollination yield and the fruit abscission rate reduced by 32.9 and 20.27, respectively[24].

In addition, the effect of using a variety of bees for pear and peach pollination was clear. Liu Jinli et al. increased the fruit yield of crown pears, emerald pears, south fruit pears, and gold pears by using concavelipped osmia for pollination by 11 to 18.4 compared with that of artificial pollination[25]. Guo Yuan et al. researched different pollination methods for pear and found that the fruit yield of bee pollination was 32.9, artificial pollination 13.05, and natural pollination only 2.83[26]. Dong Jie et al. used Italian bees and *Bombus hypocrita* to pollinate peach trees; the results showed that any difference in peach fruit yield and fruit nutritional quality of the two kinds of bee pollination was not significant and that both were significantly better than that of artificial pollination[27]. Mu Hongjie used bumblebees and bees to pollinate fruit trees; the results showed that the fruit yield of bumblebee pollination was higher than that of bee pollination, with an increase of 25.5 in the nectarine yield[28]. Means within a column followed by the different letters are extremely significant different at P 0.01 level.

From 2008 to 2015, researchers from the Horticultural Institute of the Shanxi Academy of Agricultural Sciences carried out research into bee pollination for the Red Fuji apple; the results show that bees can significantly improve the fruitsetting rate (table 2).

Table 2. Fruitsetting rate by bee and natural pollination.

Variety	Pollination pattern	Flower number	Fruit number	Fruitsetting rate ()
"Red Fuji"	Bee	1004	240	23.9
	Nature	1044	96	9.2
"Starkrimson"	Bee	1258	438	34.8
	Nature	1098	93	8.5

In addition, three, five, seven, and nine lots of bee pollination resulted in increased fruit yield to the tune of 11.1, 19, 21.7, and 25, respectively, as shown in figure 5. So the greater the number of visits made by bees to pollinate the higher the fruit yield.

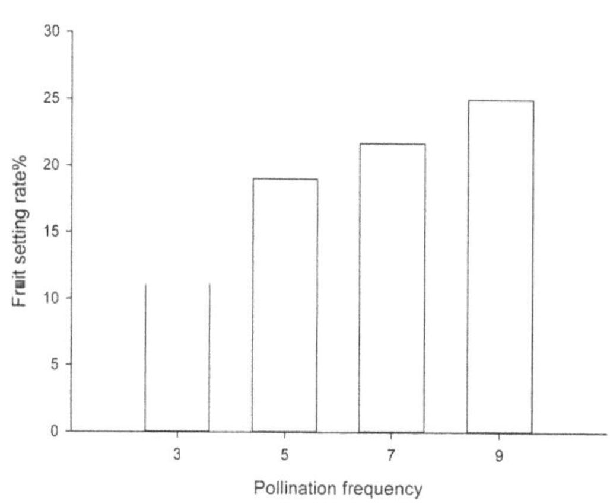

Figure 5. Fruit setting rate by number of pollination visits.

Bee pollination stimulates the growth of young fruit. The average yield of each tree after bee pollination was 69.8 kg compared with 31.9 kg of natural pollination. The fruit shape index of bee pollination and natural pollination were similar, but the coloring index of bee pollination was significantly better than the natural pollination group; the results are shown in table 3.

Table 3. Different pollination patterns.

Pollination pattern	Fruit shape index	Color index ()	Chemical quality		
			Solid content ()	Acidity ()	TSS:acid ratio
Bee	0.8282	67.33	13.531.63A	0.340.05B	40.98 A
Nature	0.8516	47.33	13.931.34A	0.430.07A	32.71 B

Notes: Fruit shape index: The ratio of longitudinal diameter to transverse diameter. Color index(Fruit number of each classExtreme value)/(Total fruit numberThe highest series)100. TSS, total soluble solids.

3.3. The fundamental technology that supports the process of bee pollination in apple and pear cultivation.

3.3.1. Configuration of pollination trees

Approximately 70% of trees being cultivated in orchards in Shanxi do not receive adequate or any pollination. We have conducted a study on the correlation between pollination partners and the cultivation of fruit trees. Some pear gardens utilize pollination branch grafting to ensure bee pollination. Studies indicate that self-incompatibility arises when pear trees possess identical Stype genotypes. Therefore, it is necessary to cultivate at least one S genotype from various varieties as pollination partners. Table 4 provides a comprehensive overview of the primary characteristics of many kinds and their compatible pollination partners.

Table 4. Configuration of main pear variety and appropriate pollination partners in Shanxi.

Main varieties	Pollination partners
Pyris pyrifolia Nak. *cuiguang*	*P. pyrifolia* Nak. *qingxiang, huanghua* *Pyris bretschneideri* Rehd. *huangguang, xinya*
P. pyrifolia Nak. *xizilv*	*P. bretschneideri* Rehd. *zaosu, hangqing* *P. bretschneideri* Rehd. *huangguang* *P. bretschneideri* Rehd. *zhongliyihao*
P. bretschneideri Rehd. *huangguang*	*jimi, P. bretschneideri* Rehd. *zhongliyihao, fengshui*
yuanhuang	*xianhuang, fengshui, huanghua, xueqing*
fengshui	*huanghua, xinshui, P. bretschneideri* Rehd. *dangshansu, P. bretschneideri* Rehd. *huangguang*

Main varieties	Pollination partners
Pyris ussriensis Max. *nanguo*	*P. bretschneideri* Rehd. *pingguo*, Bartlett Williams, *P. bretschneideri* Rehd. *ren*, *P. bretschneideri* Rehd. *ya*
xingao	*P. bretschneideri* Rehd. *ya*, *P. ussriensis* Max. *jingbai*, *P. bretschneideri* Rehd. *dangshansu*, *fengshui*
P. bretschneideri Rehd. *dangshansu*	*P. bretschneideri* Rehd. *ren*, *P. bretschneideri* Rehd. *ya*, *matihuang*, *P. bretschneideri* Rehd. *zhongliyihao*, *P. bretschneideri* Rehd. *huangguang*
P. bretschneideri Rehd. *xuehua*	*P. bretschneideri* Rehd. *ya*, *P. bretschneideri* Rehd. *zaosu*, *jimi*, *P. bretschneideri* Rehd. *huangguang*
P. bretschneideri Rehd. *ya*	*P. bretschneideri* Rehd. *dangshansu*, *P. ussriensis* Max., *jingbai*, *jinhua*
P. bretschneideri Rehd. *korla*	*P. bretschneideri* Rehd. *ya*, *P. bretschneideri* Rehd. *xuehua*, *P. bretschneideri* Rehd. *dangshansu*, *P. bretschneideri* Rehd. *pingguo*
P. bretschneideri Rehd. *hongxiangsu*	*P. bretschneideri* Rehd. *dangshansu*, *P. bretschneideri* Rehd. *xuehua*, *P. bretschneideri* Rehd. *ya*, *fengshui*

The ratio between pollination partners and the main variety can be 1:41:8; 1:6 has been found to be optimal. To ensure full pollination and prevent flowering inconsistency the main variety should be paired with two pollination partners; namely, six main varieties and one pollination variety.

In addition, we can increase the supply of pollen by means of grafting pollination branches (figure 6).

Figure 6. Central branch grafting of pear pollination partners.

3.3.2. Selection of pollinating bee species

Apis mellifera cv. The Kaqian Black Ring Bee is a cultivated variety of the Apis mellifera species. Mr. Northeast Black Bee is a subspecies of the honey bee known as Apis mellifera caucasica, also referred to as Apis mellifera cv. Honeyproplis 1 Bee, scientifically known as Apis mellifera cv. The Carpathian Bee, scientifically known as Apis mellifera cv. Heimeiyi, Apis mellifera carnica, commonly known as the Italian bee, and Apis cerana cerana Fabricius were typically employed for the sake of pollination. Nevertheless, the nine cultivars exhibit variations in their behavior patterns (figure 7), their capacity to transport pollen (figure 8), and their accumulation of pear flower powder ratio (figure 9). Hence, throughout the pollination process, it is imperative to select suitable bees based on the specific fruit tree species.

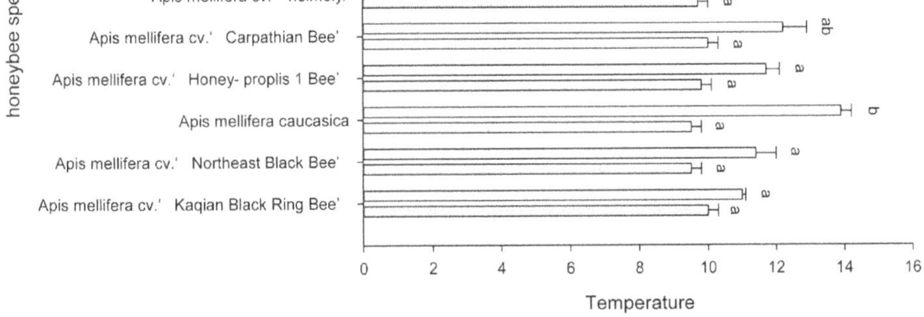

Figure 7. Temperature of outflying and backflying bee varieties.

Figure 7 shows that the outflying temperature of *Apis cerana cerana* Fabricius was 8.2°C, which was significantly lower than western bees (9.2 10°C). *Apis mellifera carnica* can fly out of the nest at 9.2°C; there were no significant differences among western bees. The average backflying temperature of all bee species was 12.1°C and the average pollencarrying temperature of *Apis mellifera* cv. Kaqian Black Ring Bee, *Apis mellifera* cv. Northeast Black Bee, *Apis mellifera* cv. Honeyproplis 1 Bee, *Apis mellifera carnica*, and *Apis cerana cerana* Fabricius was significantly lower than *Apis mellifera caucasica* and the Italian bee.

The weight of total pollen and pear pollen collected by bees in one hour was compared and analyzed (figure 8). *Apis mellifera carnica* collected more pollen than the others, and *Apis cerana cerana* Fabricius collected the least. *Apis mellifera* cv. Kaqian Black Ring Bee collected the most pear pollen in one hour and *Apis mellifera* cv. Honeyproplis 1 Bee and *Apis cerana cerana* Fabricius collected the least pear pollen. *Apis mellifera* cv. Kaqian Black Ring Bee and *Apis mellifera carnica* were much the same but they both collected more pear pollen than the others. They can collect five times more pear pollen than *Apis cerana cerana* Fabricius.

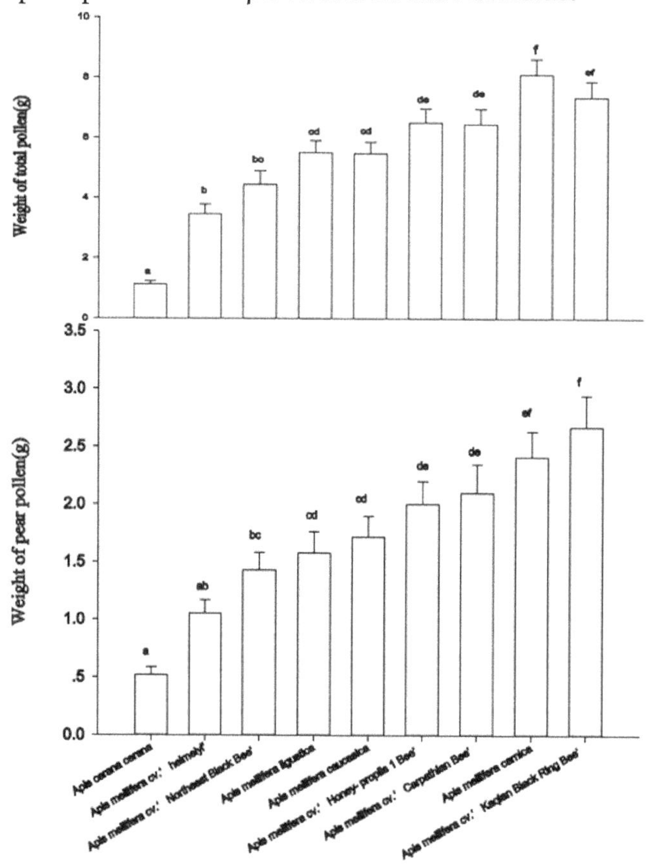

Figure 8. Total pollen and pear pollen collected by different bee varieties in one hour.

The proportion of pear pollen collected by *Apis mellifera carnica* was highest (45.2) and *Apis mellifera* cv. Kaqian Black Ring Bee was the second highest (42.4). Nevertheless, there was no significant difference among the species (figure 9).

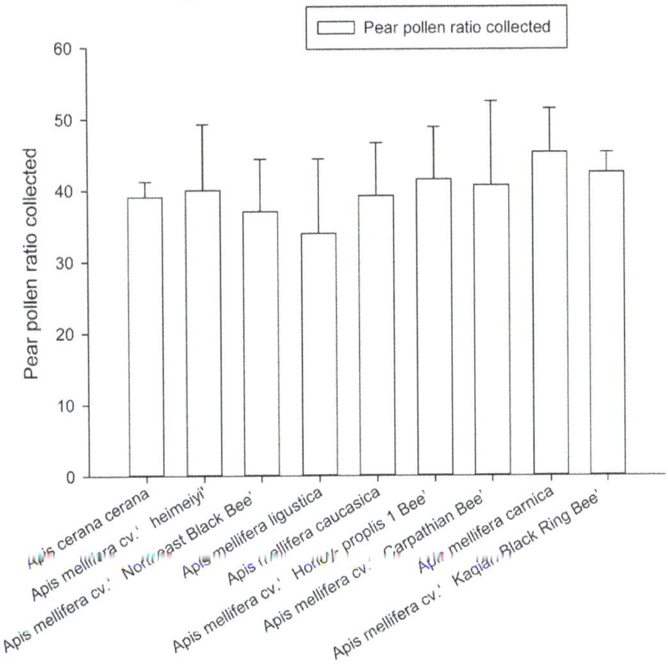

Figure 9. Pear pollen collected by different bee varieties.

Given the varying blooming periods of pear trees across China, it is necessary to have pollinating bees that excel in both pollen gathering and environmental adaptability. Pear trees exhibit early flowering in response to low temperatures, hence compelling bees to pollinate them in such conditions. Out of the nine chosen bee species, Apis mellifera cv. Kaqian Black Ring Bee and Apis mellifera carnica demonstrated superior adaptation to the environment and excelled in pollen collection, particularly when it came to pear pollen. Therefore, they can be suggested for the purpose of pollinating pear trees.

3.3.3. Control of pollination bees

Utilizing insect pollination can enhance the rate and yield of fruit production. However, it should be noted that a higher fruitbearing rate does not necessarily guarantee a better yield. Excessive fruitbearing can lead to nutritional depletion, resulting in the production of small-sized fruit and a low yield. Hence, the primary technological challenge lies in regulating the bee population to manage the fruit-setting rate.

Table 5 demonstrates a positive correlation between the number of bee kinds and several important factors: a decrease in stigma pollen subsidence, an increase in fruit-setting rate, and a higher yield. The fruit set rate is low while using one to four bee kinds. However, by increasing the number of bee varieties to six or eight, the fruit set rate might potentially increase to 9.6 or 17.9, respectively. Furthermore, the utilization of various forms of pollination by bees can also have an impact on the quality of apples, as indicated in table 6. By utilizing six different types of pollination bees, the fruit not only satisfied the necessary criteria but also possessed a pleasing taste. In summary, when the ratio of pollination partners to primary kinds is 1:4, each tree may fulfill production needs with six bees.

Table 5. Stigma pollen count, fruitsetting rate, and yield employing different numbers of varieties of pollination bees.

Number of bee varieties used	Pollen amount on stigma	Fruitsetting rate ()	Average yield/each tree (kg)
1	678.52	2.87	**65.75**
2	832.08	3.07	**65.5**
4	1,112.9	3.31	**66.5**
6	1,145.8	9.63	**79.25**
8	**1,360.4**	**17.87**	**119.25**
Notes: Experiments were conducted in net houses. The ratio between pollination partners and main varieties is 1:4.			

Table 6. Fruit quality employing different numbers of varieties of pollination bees.

Number of bee varieties used	Soluble solids ()	Acid ()	Hardiness in the sun (N/cm^2)	Hardiness in the shade (N/cm^2)	Fruit shape index	TSSacid ratio
1	15.0527bc	0.302d	9.154b	9.24b	86.44	49.79
2	14.8423c	0.318cd	10.809a	10.72a	87.81	46.7
4	15.9480ab	0.3906b	10.509a	10.23a	86.96	40.82
6	16.6893a	0.502a	10.46a	10.31a	88.49	33.25
8	15.6733bc	0.375bc	9.485b	8.73b	85.65	41.78
Note: TSS, total soluble solids. Means within a column followed by the different letters are significantly different at P 0.05 level.						

When the ratio between the main variety of “Red Fuji” apples and pollination partners of “Qinguan” is 20:1, the fruit yield, fruit shape index, seed number, and deformity fruit rate of each tree employing 6, 12, and 18 varieties of bees are, respectively, shown in figure 10, table 7, and table 8. Notes: Means within a column followed by the same letter are not significantly different.

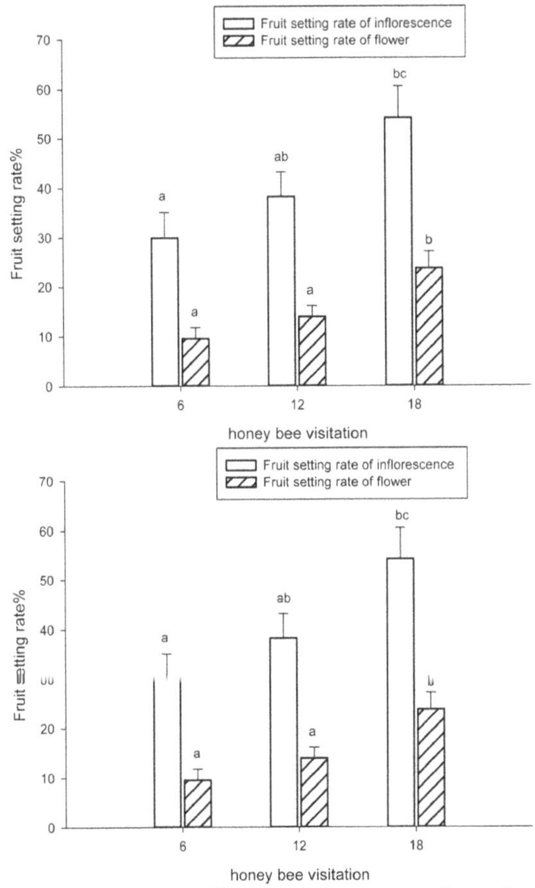

Figure 10. Number of varieties of pollination bees employed and fruitsetting rate.

Table 7. Relation between number of varieties of pollination bees employed and fruit shape index.

Number of bee varieties used (each tree)	Average longitudinal diameter (mm)	Average diameter (mm)	Fruit shape index
6	66.41	75.26	0.91110.0103a
12	68.2	75.21	0.90760.0094a
18	67.65	74.28	0.88210.0121a

Table 8. Relation between number of varieties of pollination bees and fruit seed number and irregular fruit rate.

Number of bee varieties used (each tree)	Total number of seeds	Plump seed	Plumpness ()	Irregular fruit rate ()
6	140	134	95.7	56a
12	247	243	98.38	26b
18	290	288	99.31	18c

In the absence of pollination partners the number of bee varieties used for pollination and the fruitsetting rate are shown in figure 11.

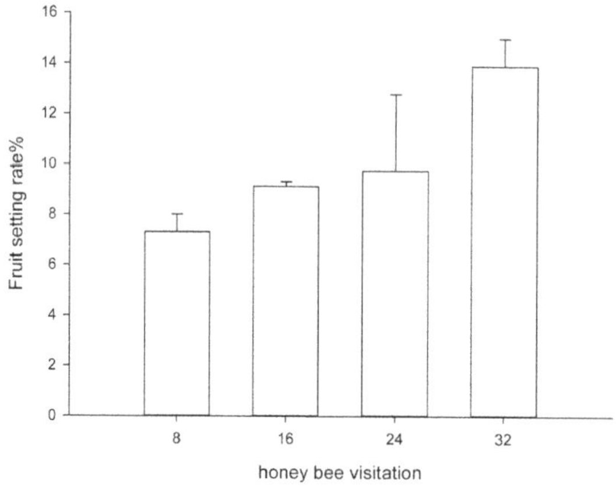

Figure 11. Fruit setting rate of "Red Fuji" apple trees by bee pollination in the absence of pollination partners.

The arrangement of pollination partners is a crucial determinant of the quantity of bee species required for pollination. An optimal arrangement will enable the utilization of six different types of pollination bees, which will be sufficient to ensure the fulfillment of production needs. If the current configuration does not have enough pollination partners, it will be essential to augment the number of bee kinds to a minimum of 12 for the purpose of pollination. To reach a fruitsetting rate of 9.12, it is necessary to have 16 different types of bees when there are no other bees available for pollination.

3.3.4. Scale of the Pollination Apiary

The proximity of buzzers to fruit trees had a notable impact on the process of pollination. Tables 9 and 10 display the data on the foraging behavior of bees and the percentage of fruiting at various distances. The data was collected from 20 colonies.

Table 9. Number of foraging bees and fruiting percentage at different distances using 20 colonies in 2009.

Distance (m)	Number of foraging bees	Flower number	Fruit number	Fruiting percentage
0	354	715	138	19.30
50	246	610	94	15.41
100	227	764	97	12.70
150	194	935	112	11.98
200	86	1,825	126	6.90

Notes: 20 colony treatment involved setting up 5 survey spots, 1 every 50 m from the colony out to 300 m. At every spot one apple tree of consistent variety, tree potential, onyear yield (high), and offyear yield (low) was selected. At every spot foraging bees were counted for 45 minutes and the fruiting percentage after 15 days was calculated.

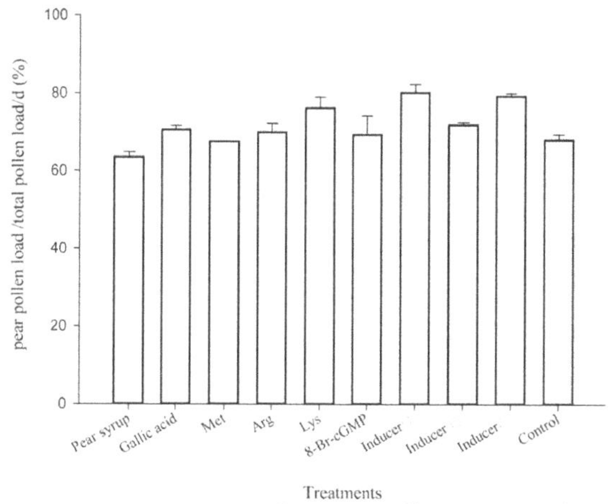

Figure 13. Percentage of pear pollen in total pollen.

Cameras were installed at the entrance to the hive, worker bees returning to the hive were recorded for 5 minutes every hour. The percentage of foraging bees on pear flowers out of total foraging bees is shown in figure 14. The results showed that treatment groups fed pear syrup, gallic acid, and methionine were significantly lower than the control group, whereas the group fed arginine and 8BrcGMP showed no significant difference from the control group. The groups fed lysine (85.81), hanging attractant I (86.74), attractant II (87.27), and attractant III (85.67) were significantly higher than the control group. The percentage of foraging bees improved by 3 to 4.5 after application of attractants I, II, and III.

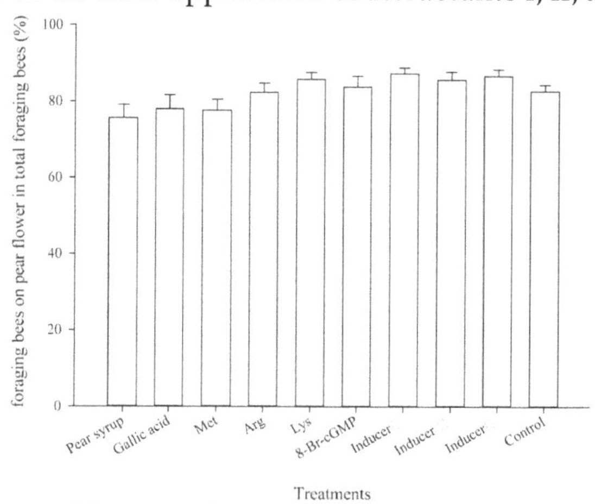

Figure 14. Percentage of foraging bees on pear flowers out of total foraging bees.

Based on the aforementioned results, it is evident that providing bees with lysine yields the most favorable outcome among all the feeding methods. This leads to an increase in both the proportion of pear pollen carried by bees and the number of bees foraging on pear flowers. The efficacy of using hanging attractants in the hives surpasses that of feeding treatments. The utilization of Hanging attractant II in the hives led to the most significant proportion of pear pollen load, while the

Table 10. Fruiting percentage at different distances using 20 colonies in 2010.

Distance (m)	Flower number	Fruit number	Fruiting percentage	Pollen amount on stigma
0	1,723	286	16.6	6,050
50	1,843	304	16.49	5,450
100	3,090	417	13.5	6,889
150	1,738	228	13.12	5,850
200	1,790	202	11.28	6,050

With increase of the distance between fruit trees and bee colonies the number of foraging bees gradually reduced. Bee pollination at a distance of 150 m between bee colonies and fruit trees was found to give the ideal fruiting percentage; therefore, the bee pollination effective radius was 150 m when 20 colonies are employed.

With increase of the distance in the 0 to 200m range using 20 colonies the fruiting percentage decreased from 16.60 to 11.28; however, neither the trend nor change in the pollen count on the stigma were obvious. Despite there being more foraging bees and the fruiting percentage increasing with decrease in the distance from the colony within the 0 to 200m range, 2010 was an offyear with low apple tree yield, reduced flower total quantity, and enlarged bee gather distance.

The number of foraging bees and fruiting percentage at different distances using 50 colonies are shown in tables 11 and 12.

Table 11. Foraging bees and fruiting percentage at different distances using 50 colonies in 2009.

Distance (m)	Number of foraging bees	Flower number	Fruit number	Fruiting percentage
0	907	1,070	339	31.68
50	593	1,775	236	13.30
100	600	1,250	140	11.20
150	342	1,865	208	11.15
200	262	1,400	148	10.57
250	241	1,255	70	5.58
300	189	1,580	97	6.14

Notes: 50colony treatment involved setting up 7 survey spots, 1 every 50 m from the colony out to 300 m. At every spot one apple tree of consistent variety, tree potential, and onyear and offyear yields was selected. At every spot foraging bees were counted for 45 minutes and the fruiting percentage after 15 days was calculated.

Table 12. Fruiting percentage at different distances using 50 colonies in 2010.

Distance (m)	Flower number	Fruit number	Fruiting percentage	Pollen amount
0	1,561	309	19.8	5,400
50	592	93	15.71	5,050
100	1,200	135	11.25	5,167
200	1,425	203	14.25	4,800
250	1,097	215	19.60	4,000
300	1,373	202	14.71	4,000
Note: There were no 150m data because tree vigor was poorer.				

When 50 colonies were used, the fruiting percentage declined from 31.7% to 6.1% as the distance increased from 0 to 300m in 2009. At a distance of 200m, the fruiting percentage was 10.2%, and at a distance of 250m, it was 5.6%. Therefore, the effective radius for bee pollination is 200m when 50 colonies are used. Apple pollen levels on stigmas exhibited a general decrease as the distance increased in 2010. This data indicates that the number of bees searching for food decreased as the distance increased. However, there was no noticeable change in the percentage of fruiting plants.

The effective radius of bee pollination was 150 m when there were 20 colonies, and it increased to 200 m when there were 50 colonies. When apple trees produced one-year yields, production practices selected 50 colonies for bee pollination. During years when apple trees produced lower yields, the effective radius of bee pollination was greater compared to years when apple trees produced higher yields. In order to enhance pollination, 20 colonies of bees were selected for the production practice.

3.3.5. Technology behind getting bees to visit fruit trees

Pear trees have low pollinator attractiveness. When pear trees are surrounded by blooming plants like rape, dandelion, and paulownia, insects that search for food seldom land on the pear trees[29]. The issue of artificially stimulating bees to pollinate fruit trees instead of more preferable plants necessitates a resolution. Fruit trees often experience early blooming in the year, with a relatively brief flowering duration. over instance, apple trees exhibit flowering over a span of 10 to 15 days, whereas pear trees have a shorter flowering period of approximately 7 to 10 days. Another challenge is in stimulating the foraging motivation of the swarm.

In an effort to improve the foraging enthusiasm of bees, our team studied foraging behavior after using attractants. *Apis mellifera* 30 hives with six combs every hive had a consistent colony structure. The test involved 10 treatments and

3 colonies. Three treatments, respectively, used attractant I, II, and III, which our team prepared. Nine treatments involved feeding six compounds to bees: 1mM methionine, 1mM lysine, 1mM arginine, 1mM gallic acid, 500μM 8BrcGMP, and pure syrup as a control treatment. Attractants were startfed to *Apis mellifera* once every evening before the pear blossom appeared and then every 2 days until the end of the flowering. Pear flower load after sorting from total pollen was weighed, number of foraging bees on pear flowers were recorded, and percentage pear flower load and foraging bee number on pear flowers were calculated.

The weight of pear pollen load collected in a day is shown in figure 12. All treatments enhanced the foraging ability of bees for pear pollen. The foraging effect of hanging attractant I in hives (77.56 1.59 g/group) was significantly higher than other treatments. Those treatments that involved feeding the bees Arg (62.05 / 2.01 g), Lys (62.2 2.3 g), 8BrcGMP (64.45 4.55 g) and hanging attractant II in hives (64.20 2.72 g) were better than the control treatment (49.11 1.03); other treatments showed no significant difference from the control treatment.

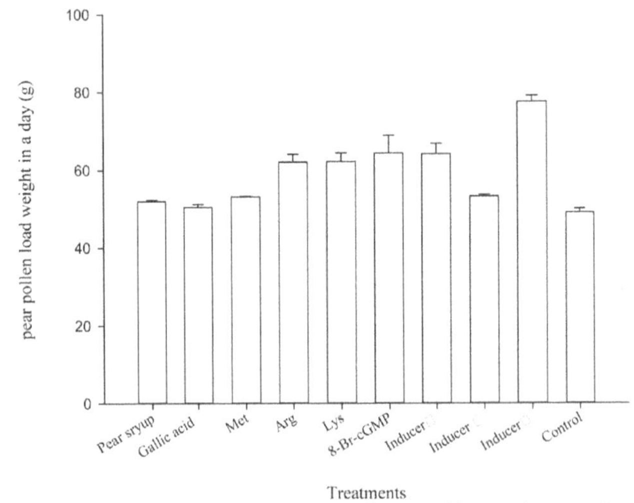

Figure 12. Weight of pear pollen collected in a day.

Pollen from pear trees was separated from the whole pollen sample. Figure 13 displays the proportion of pear pollen in the overall pollen collection for a given day. The findings indicate that the treatment groups administered with pear syrup and methionine did not exhibit any notable disparity compared to the control group. Conversely, the remaining seven groups demonstrated greater values than the control group. The levels of lysine (76.3), hanging attractant I in the hives (79.3), and hanging attractant II in the hives (80.2) were considerably greater than those in the other treatment groups (P < 0.05). The most effective method was to hang attractant II in the hives, which resulted in the highest proportion of pear pollen. The next best option was to hang attractant I in hives that were linked w the lysine groups.

Table 10. Fruiting percentage at different distances using 20 colonies in 2010.

Distance (m)	Flower number	Fruit number	Fruiting percentage	Pollen amount on stigma
0	1,723	286	16.6	6,050
50	1,843	304	16.49	5,450
100	3,090	417	13.5	6,889
150	1,738	228	13.12	5,850
200	1,790	202	11.28	6,050

With increase of the distance between fruit trees and bee colonies the number of foraging bees gradually reduced. Bee pollination at a distance of 150 m between bee colonies and fruit trees was found to give the ideal fruiting percentage; therefore, the bee pollination effective radius was 150 m when 20 colonies are employed.

With increase of the distance in the 0 to 200m range using 20 colonies the fruiting percentage decreased from 16.60 to 11.28; however, neither the trend nor change in the pollen count on the stigma were obvious. Despite there being more foraging bees and the fruiting percentage increasing with decrease in the distance from the colony within the 0 to 200m range, 2010 was an offyear with low apple tree yield, reduced flower total quantity, and enlarged bee gather distance.

The number of foraging bees and fruiting percentage at different distances using 50 colonies are shown in tables 11 and 12.

Table 11. Foraging bees and fruiting percentage at different distances using 50 colonies in 2009.

Distance (m)	Number of foraging bees	Flower number	Fruit number	Fruiting percentage
0	907	1,070	339	31.68
50	593	1,775	236	13.30
100	600	1,250	140	11.20
150	342	1,865	208	11.15
200	262	1,400	148	10.57
250	241	1,255	70	5.58
300	189	1,580	97	6.14

Notes: 50colony treatment involved setting up 7 survey spots, 1 every 50 m from the colony out to 300 m. At every spot one apple tree of consistent variety, tree potential, and onyear and offyear yields was selected. At every spot foraging bees were counted for 45 minutes and the fruiting percentage after 15 days was calculated.

Table 12. Fruiting percentage at different distances using 50 colonies in 2010.

Distance (m)	Flower number	Fruit number	Fruiting percentage	Pollen amount
0	1,561	309	19.8	5,400
50	592	93	15.71	5,050
100	1,200	135	11.25	5,167
200	1,425	203	14.25	4,800
250	1,097	215	19.60	4,000
300	1,373	202	14.71	4,000
Note: There were no 150m data because tree vigor was poorer.				

When 50 colonies were used, the fruiting percentage declined from 31.7% to 6.1% as the distance increased from 0 to 300m in 2009. At a distance of 200m, the fruiting percentage was 10.2%, and at a distance of 250m, it was 5.6%. Therefore, the effective radius for bee pollination is 200m when 50 colonies are used. Apple pollen levels on stigmas exhibited a general decrease as the distance increased in 2010. This data indicates that the number of bees searching for food decreased as the distance increased. However, there was no noticeable change in the percentage of fruiting plants.

The effective radius of bee pollination was 150 m when there were 20 colonies, and it increased to 200 m when there were 50 colonies. When apple trees produced one-year yields, production practices selected 50 colonies for bee pollination. During years when apple trees produced lower yields, the effective radius of bee pollination was greater compared to years when apple trees produced higher yields. In order to enhance pollination, 20 colonies of bees were selected for the production practice.

3.3.5. Technology behind getting bees to visit fruit trees

Pear trees have low pollinator attractiveness. When pear trees are surrounded by blooming plants like rape, dandelion, and paulownia, insects that search for food seldom land on the pear trees[29]. The issue of artificially stimulating bees to pollinate fruit trees instead of more preferable plants necessitates a resolution. Fruit trees often experience early blooming in the year, with a relatively brief flowering duration. over instance, apple trees exhibit flowering over a span of 10 to 15 days, whereas pear trees have a shorter flowering period of approximately 7 to 10 days. Another challenge is in stimulating the foraging motivation of the swarm.

In an effort to improve the foraging enthusiasm of bees, our team studied foraging behavior after using attractants. *Apis mellifera* 30 hives with six combs every hive had a consistent colony structure. The test involved 10 treatments and

3 colonies. Three treatments, respectively, used attractant I, II, and III, which our team prepared. Nine treatments involved feeding six compounds to bees: 1mM methionine, 1mM lysine, 1mM arginine, 1mM gallic acid, 500μM 8BrcGMP, and pure syrup as a control treatment. Attractants were startfed to *Apis mellifera* once every evening before the pear blossom appeared and then every 2 days until the end of the flowering. Pear flower load after sorting from total pollen was weighed, number of foraging bees on pear flowers were recorded, and percentage pear flower load and foraging bee number on pear flowers were calculated.

The weight of pear pollen load collected in a day is shown in figure 12. All treatments enhanced the foraging ability of bees for pear pollen. The foraging effect of hanging attractant I in hives (77.56 1.59 g/group) was significantly higher than other treatments. Those treatments that involved feeding the bees Arg (62.05 / 2.01 g), Lys (62.2 2.3 g), 8BrcGMP (64.45 4.55 g) and hanging attractant II in hives (64.20 2.72 g) were better than the control treatment (49.11 1.03); other treatments showed no significant difference from the control treatment.

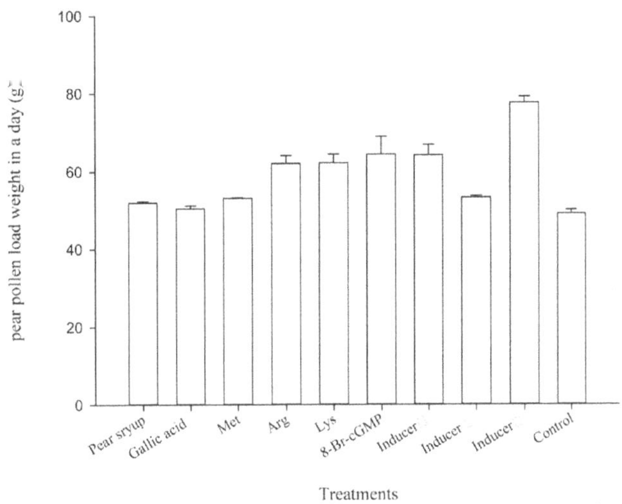

Figure 12. Weight of pear pollen collected in a day.

Pollen from pear trees was separated from the whole pollen sample. Figure 13 displays the proportion of pear pollen in the overall pollen collection for a given day. The findings indicate that the treatment groups administered with pear syrup and methionine did not exhibit any notable disparity compared to the control group. Conversely, the remaining seven groups demonstrated greater values than the control group. The levels of lysine (76.3), hanging attractant I in the hives (79.3), and hanging attractant II in the hives (80.2) were considerably greater than those in the other treatment groups ($P < 0.05$). The most effective method was to hang attractant II in the hives, which resulted in the highest proportion of pear pollen. The next best option was to hang attractant I in hives that were linked with the lysine groups.

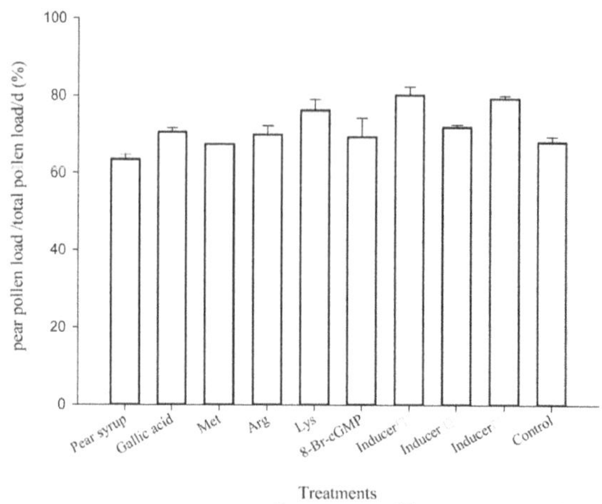

Figure 13. Percentage of pear pollen in total pollen.

Cameras were installed at the entrance to the hive, worker bees returning to the hive were recorded for 5 minutes every hour. The percentage of foraging bees on pear flowers out of total foraging bees is shown in figure 14. The results showed that treatment groups fed pear syrup, gallic acid, and methionine were significantly lower than the control group, whereas the group fed arginine and 8BrcGMP showed no significant difference from the control group. The groups fed lysine (85.81), hanging attractant I (86.74), attractant II (87.27), and attractant III (85.67) were significantly higher than the control group. The percentage of foraging bees improved by 3 to 4.5 after application of attractants I, II, and III.

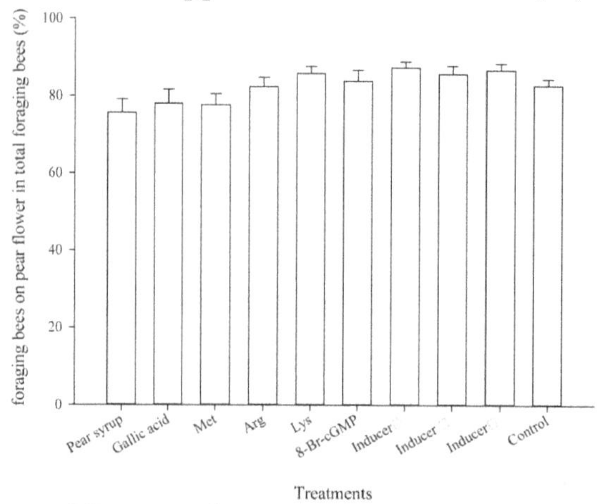

Figure 14. Percentage of foraging bees on pear flowers out of total foraging bees.

Based on the aforementioned results, it is evident that providing bees with lysine yields the most favorable outcome among all the feeding methods. This leads to an increase in both the proportion of pear pollen carried by bees and the number of bees foraging on pear flowers. The efficacy of using hanging attractants in the hives surpasses that of feeding treatments. The utilization of Hanging attractant II in the hives led to the most significant proportion of pear pollen load, while the utilization of Hanging attractant I in the hives resulted in the highest quantity of pear pollen by weight. Both treatments can significantly enhance foraging behavior.

3.3.6. Technology underlying bees carrying pollinizer pollen

Due to the absence of pollinizers in certain pear orchards, we have devised a specialized device known as a beecarrying powder device (see to figure 15). This device is positioned near the hive entrance, where fresh pollen is placed in the upper compartment. As bees exit the hive, pollen will gradually seep out from the lower half of the device. Bees that transport pollen from pollinizer plants will fertilize and spread it to the most important cultivated varieties.

Figure 15. Device for bees to carry pollen.

4. INDUSTRIALIZATION OF FRUIT TREE POLLINATION

4.1. Professional bee breeding for pollination

Apis mellifera and Apis cerana cerana are the primary bees used for pollination, particularly for the production of bee products such as honey. They are occasionally utilized for pollinating fruit trees as well. Nevertheless, they are not optimal for pollination purposes[30]. During the 1990s, Chinese researchers achieved a significant advancement in the artificial breeding of wild bumblebees. They successfully developed the necessary technology to breed bumblebees indoors and successfully domesticated six different species of bumblebees. Additionally, they established multiple production bases that could be gradually utilized for orchard pollination purposes[31]. Researchers have recently produced and tamed osmia bees, stingless bees, and andrenids to address the practical development needs in agriculture. We anticipate that these technologies will have a significant impact on fruit tree pollination in the future.

4.2 Further research is required to explore the technique of induced bee pollination

Scientists have established a unique group dedicated just to pollinating fruit trees, created a method of pollination that does not rely on a queen bee, and resolved the issue of bees colliding with greenhouses at agriculture facilities. Despite significant progress in bee pollination, some unresolved technical challenges persist. These include finding solutions to encourage bees to visit fruit trees that they are not naturally inclined towards and enhancing their pollination capabilities. The answers to these inquiries entail examining the associations between plants and bees, as well as the connection between the arrangement of fruit trees and the dispersion of foraging bees in space.

4.3. Pollination professional development is slow

Currently, certain regions in China have implemented bee pollination intermediary service agencies, bee industry cooperatives, pollination professional firms, and a limited number of matching pollination intermediary services. Nevertheless, these groups have not been successful in delivering the essential market supply and demand data, technical training, or establishing a bee pollination service and related policies. There are several factors contributing to this phenomenon: The Chinese industry is characterized by a limited scale, a scarcity of specialized enterprises, a poor level of structure, and challenges in establishing a pollination network.

REFERENCES

1. Zhang YX (Ed.) General fruit cultivation. Beijing: China Agriculture Press; 2014.
2. State Statistics Bureau, (Ed.) Statistical year book of rural China in 2012. Beijing: Chinese statistics press; 2012.
3. Available from: http://faostat.fao.org/browse/QQC/E
4. Guo GN, Yan ZL, Zhang HT, et al. Current situation of production of early and mid season apple cultivars in China and progress in breeding for early and mid season apple cultivars. Journal of Fruit Science. 2009;26(6):871–877.
5. Chen CG, Zhao DY, Xuan JH. The suggestion to apple varieties development in our country. Northern Fruits. 2015;xx(3):52–53. DOI: 10.16376/j.cnki.bfgs.2015.03.027
6. Ou XQ, Zhou Y. The analysis of concept of plants self-pollination and cross pollination. Biology report. 2007;42(11):23–25.
7. Roubik DW. Pollination of cultivated plants in the tropics. FAO Agricultural Services Bulletin. 1995;118:1–196.

8. Roubik DW (Ed.) Diversity in the real world: Tropical forests as pollinator reserves. In: Bees and the Environment (ed. Mardan M). Malaysia: South Bound Press; 1995.

9. Shen JS, Wu WQ, Song HL, et al. The study of pollinating insects and their flower-visiting behavior on Malus pumila Mill. Journal of Environmental Entomology. 2015;37(4):720–725. DOI: 10. 3969 /j. issn. 1674-0858. 2015. 04. 4

10. Lu YG, Qi WZ, Zhang SW, et al. Observational study on the insect visiting apple flowers in the central loess plateau petitions. Apiculture of China. 2012;63:27–29.

11. Yang LL, Wu YR. The preliminary research on the kiwi pollinator. China apiculture. 1990;(3):2–4.

12. Zhang YY, Wu WQ, Ma WH. The Investigation of Cherry Pollinator. Chinese Agricultural Science Bulletin. 2012;28(25):272–276.

13. Li J, Zhu CD, Wang FH, et al. Current research on the status of wild bees and their pollination roles. Biodiversity Science. 2007;15(6):687–692. DOI: 10.1360/biodiv.070080

14. Xie ZH, Xu HL, Yang P. Notes on monitoring, assessing and conserving pollinator biodiversity. Chinese Journal of Applied Entomology. 2011;48(3):746–752.

15. Huang DY, Hao JS, Yu JF, et al. Impact of Human Activities on Pollination of Economic Crops. Nonwood Forest Research. 2008;26(4):114-120.

16. Steffan-Dewenter I, Potts SG, Packer L. Pollinator diversity and crop pollination services are at risk. Trends in Ecology and Evolution. 2005;20:651.

17. Kearns CA, Inouye DW, Waser NM. Endangered mutualisms: The conservation of plant–pollinator interactions. Annual Review of Ecology and Systematics. 1998;29:83–112.

18. Wu WQ, Guo Y, Shen JS, et al. Present situation investigation of bee pollination for pear. Apiculture of China. 2011;62:40–44.

19. Li MH, Cong B, Li JP, et al. Mason Bee and Its Application on Fruit Tree Pollination. Journal of Jilin Agricultural University. 2004;26(4):422–425.

20. Zhang GQ, Tian ZZ, Shen RM, et al. The effect of Apis mellifera pollination for Fuji and golden apple. Agricultural science technology and information. 2013;xx(1):23–25. DOI:10.15979/j.cnki.cn62-1057/s.2013.01.025

21. He WZ, Zhou WR. The effect of Osmia excavata Alfken, Apis mellifera and artificial pollination for apple. Apiculture of China. 2009;60(11):9–11.

22. Lou DL, Jiang FT, Wang SQ. Study on for [sic] Apple Bee Pollination. Apiculture of China. 2013;64:35–37.

23. Zhang ZH, Yang PY, Li P, et al. The integrated application and demonstration effect of bee pollination and green early prevention and control production technology. China Plant Protection. 2015;35(4):43–45.
24. Yuan F, Wei YP, Zhang YL. Research on the result of pollination of apple by Osmia bees. Acta Agriculturae Boreali-occidentalis Sinica. 1992;1(3):93–96.
25. Liu JL, Cui LX, Le WQ. The pollination effect of Osmia excavata Alfken for different varieties of pear tree. Hebei Fruits. 2014;xx(1):44 –45.
26. Guo Y, Shao YQ, Ma WH, et al. Study on Physiological Activity and Pollination Characteristic of Stigma and Pollen of 'Dangshansuli'. Chinese Agricultural Science Bulletin. 2013;29(16):110 –114.
27. Dong J, An JD, Huang JX, et al. Effects of pollination by different bees on peach fruit development and quality under greenhouse conditions. Chinese Journal of Eco-Agriculture. 2011;19(4):836-842.
28. Mu HJ, Sun ZP, Zhang SH, et al. The test of bumblebee pollination for facilities fruit trees. Hebei Fruits. 2006;xx(6):6–7.
29. Teng TZ, Huo YG, Li Y. The relationship of pear fertilization and bee pollination. Apiculture of China. 2011;ZA:19–21.
30. Peng WJ, Wu J, An JD, et al. Comparative studies on pollination biologies of Bombus lucorum and Apis mellifera ligustica in Katy apricot greenhouse. Journal of Fujian Agriculture and Forestry University (Natural Science Edition). 2007;36(3):302–306.
31. Wu J, Shao YQ (Eds.) Wonderful efficient crop production technology—honeybee pollination. Beijing: China Agriculture Press; 2011.

07 BEEKEEPING IN JALISCO, MÉXICO

1. INTRODUCTION: AN OVERVIEW

Apiculture is a field of production that is conducted using a diverse range of technologies and involves both vertical and horizontal integration of the production process. Agriculture is a significant sector in Mexico that encompasses various aspects such as food production, economic growth, social development, and ecological sustainability. It has flourished in different regions of the country, primarily driven by small and medium-sized producers who have a significant presence in both the international market and local consumption. In the 1990s, the average per capita consumption of agricultural products was 190 grams, which rose to 320 grams by 2010. The reason for this rise is its utilization as a primary ingredient in the production of various food items such as yogurt, cereals, sweets, pastries, and cosmetic goods.[1] Mexico ranked sixth globally in honey production in 2010, with 1.8 million beehives yielding 56,883 tons annually. Additionally, it stood as the third largest honey exporter, shipping 25,000 tons to the European market that year. This thriving apiculture industry played a significant role in generating foreign currency, positioning it among the top three contributors in the national livestock sector.[3] Annually, a staggering quantity of over 2,400 tons of beeswax and nearly 8 tons of royal jelly are generated.[4] Apiculture provides direct advantages to 400,000 individuals involved in the beekeeping production chain, including those engaged in the construction of beekeeping equipment and the processing and selling of honey and other bee products. Beekeeping not only contributes to the pollination of agricultural crops, which is valued at around 2 billion dollars annually, but it also plays a crucial role in preserving ecological equilibrium in many environments by pollinating wild plants.[5] Jalisco, a state in Mexico, is a significant producer of honey. It has a total of 157,827 hives, which produce an average of 5,698 tons of honey per year from 2005 to 2009. Jalisco holds a 10% market share and ranks third nationally, behind Yucatán with nearly 10,000 tons (15%) and Campeche with 7,500 tons (12.9%). There are almost 1,000 beekeepers in the state, with half of them located in the southern and southeastern districts, which are the primary areas for beekeeping due to their fertile environment. Beekeeping is primarily practiced as a secondary occupation to agriculture and cattle.The list [1,6,7]

Over the past few decades, the beekeeping industry has seen significant transformations due to urbanization, globalization, and population increase, resulting in the emergence of a new ecosystem.The given text is "[8,9]". Efforts have been made to enhance production, enhance the variety of the final product, and experiment with new organizational strategies, resulting in the emergence of novel commercial dynamics and approaches to entering the global market.[1] The government has prioritized efforts to encourage productive restructuring, diversify conventional crops, evaluate technology, and foster the development of infrastructure and technical innovation.The numbers 9 and 10 are enclosed in square brackets [9, 10].

Multiple studies have highlighted the various challenges that impact national beekeeping, including the presence of Africanized bees, global climate change (which encompasses factors like unpredictable rainy seasons, droughts, extreme heat, and freezing temperatures), as well as inadequate training and organization of beekeepers. Additionally, diseases such as varroa and foulbrood contribute to these issues. Intermediaries and market rivalry at the global level have also played a role in fostering a concerning lack of stability.The list [3,6,11]

Honey had been experiencing significant annual sales growth until 2007. [12] The increasing global honey market is driven by an optimistic prognosis, as the existing supply fails to meet the whole demand.The user's text is "[13]". Nevertheless, marketing poses as another fundamental challenge within this production industry. To classify product creation as a major activity rather than a supplementary source of income, it is necessary to make modifications to the collection method, presentation, and business dynamic. Typically, honey in Mexico is regarded as a secondary commodity, with only a small number of producers and companies dedicating resources to research, development, preservation, and enhancing its quality. Additionally, there is limited focus on diversifying sales methods and implementing effective marketing strategies and distribution channels.The text is enclosed in the [14] tags.

In the realm of Mexican apiculture, over 75% of beekeepers are impoverished farmers who view apiculture as a means of augmenting their earnings. On average, they possess less than 100 beehives, a figure that is diminishing due to the aforementioned issues. The small producers' methods of operation do not adhere to business logistics, resulting in challenges in obtaining accurate data regarding their income. They do not maintain any records of production, expenditures, or revenue.

Considering the economic and social significance of apiculture in the southern and southeastern regions of Jalisco, along with the ongoing changes in the industry, it is crucial to analyze the characteristics of beekeepers and innovation agents, taking into consideration socioeconomic, technological, and productive factors. The objective of this study was to identify the socioeconomic factors that have a significant impact on the beekeeping process and to provide an accurate description of the current technological development of beekeepers in the south and southeast regions of Jalisco. The purpose of this research is to obtain an updated and objective understanding of the apiculture sector, which can serve as a reference point and a crucial decision-making tool for government support programs aimed at benefiting beekeepers.

2. MATERIALS AND METHODS

The documentary material was gathered from secondary sources to establish a baseline understanding of the production and commercial statistical patterns within the apiculture production chain, as well as the methodological framework.

The study design is characterized by its exploratory nature and reliance on quantitative methods. Exploratory investigation is employed to delineate the study problem and its context by scrutinizing secondary data. The study employed a descriptive and cross-sectional quantitative approach, utilizing a person-to-person nominal scale survey and a single sample method. The main data was collected by a survey utilizing a well-organized questionnaire. The user's text is "[15,16]". We conducted an analysis of various factors including demographic, social, technological, and economic variables. These variables encompassed questions related to gender, age, frequency of hive inspections, varroa treatments, price of honey per kilo, apiculture products, type of extraction equipment, labeling and marketing strategies for honey, main diseases and their treatments, type of feed and its frequency, extraction equipment, and production costs, among others.The user's text is "[17]".

A pilot survey was applied beforehand to 30 people to make adjustments to the final questionnaire. A stratified sampling of beekeepers was made in six strata: 1 to 25 hives, 25 to 50 hives, 50 to 100 hives, 100 to 500 hives, 500 to 1,000 hives, and over 1,000 hives. From a population of 1,000 beekeepers in Jalisco, 50% live in the study zones, resulting in a population of 500 beekeepers. The final sampling was of 183 beekeepers surveyed, with a 95% confidence level, 3% accuracy, 5% participation, and sample size adjusted to 15% losses. For the data analysis we used a descriptive and quantitative method to identify, understand, correlate, and prove the hypothesis of the study. The acquired information was analyzed using the SPSS version 19® statistical software. We utilized frequency analysis, ANOVA (Waller-Duncan), and contingency tables ($\chi 2$) to determine if there exists a correlation between the social, economic, and technological profile variables within the apiculture production chain.

3. RESULTS AND DISCUSSION

3.1. Demographic aspects

In the research region, a significant proportion (59%) of beekeepers handle less than 100 hives. However, it was noted that a few beekeepers have more than 1,000 hives (see Figure 1). In the research zone, the mean age of male beekeepers was 47 years, whereas the mean age of female beekeepers was 45 years. Merely 3% of individuals were below the age of 20, and the beekeepers who possessed the greatest number of hives were also the most advanced in age, indicating a noticeable absence of enthusiasm for beekeeping among the younger generation. The average age in the Yucatán peninsula is 47 years, which is different from the reported average age of 43 years in Michoacán according to a survey conducted in 2004.The given text is the list [19, 20]. Conversely, the average age of the island of Tenerife, Spain, is shown to be 59 years.The user's text is "[21]". Age is a crucial determinant when it comes to the current and future proficiency of beekeepers in managing their bees. Older beekeepers tend to be less open to adopting new production or management approaches, preferring to stick to their conventional practices. Similarly, collaborating on projects with beekeepers who are under the age of 25-30 can lead to instability due to temporary or permanent relocation caused by a scarcity of job opportunities in the field or possibly due to educational pursuits.The text is enclosed in the tags [22].

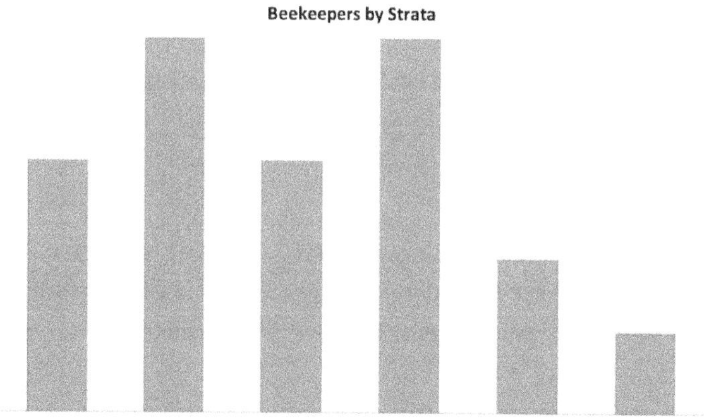

Figure 1. Distribution of beekeepers by strata. (Livestock Research Laboratory, Department of Agriculture Production, CUCSUR, University of Guadalajara).

3.2. Location of apiaries

Regarding access to land for beekeeping, there is a notable disparity ($p = 0.046, \chi 2$) in which 61.2% of beekeepers lease the property where they keep their beehives, while the remaining beekeepers utilize tiny plots of land and, to a lesser extent, communal or ejido land. Beekeepers who manage 101 to 500 hives are more inclined to lease their premises, mirroring the situation observed in Michoacán towns where the majority of beekeepers do not possess the land on which they establish their apiaries..[23] This situation limits apiculture development as producers must pay, either in cash or kind, for the lease of the lands, and also limits the assurance of the availability of the space to maintain the apiaries; this, in addition to competition for better spaces not only among beekeepers but also farmers and other branches of livestock. In Turkey, 90.59% of beekeepers have their apiaries installed on private property.[20]

Some 60% of beekeepers installed their apiaries in the municipalities where they were born and the rest look elsewhere for suitable flowering spaces, showing a significant difference ($p = 0.000$, χ^2) where beekeepers with more than 100 hives have greater mobility in search of better yields. They also mention a wide saturation of hives in their municipalities, this being another of the main issues raised within the beekeeping production system. This is related to the average distance of 25 km they have to travel to inspect the apiaries (in a range from 1 to 200 km), where a significant difference ($p = 0.000$, χ^2) was observed, beekeepers with less than 50 hives traveling less than 10 kilometers to install their apiaries, while those with over 500 hives travel distances in excess of 60 kilometers. Such a situation is unique to this region; a study in Chile describes a high concentration of apiaries in certain zones,[22] a fact that goes against the environmental management requirements for good farming practices. A different situation exists in Yucatán, where close to 50% of beekeepers travel more than 10 kilometers to reach the apiary, while 22.2% travel less than 2 kilometers, which leads to strong competition among the bees to obtain food, since 88.9% have apiaries at a distance of less than the recommended 3 kilometers.[24]

3.3. Months of honey production

All beekeepers, without exception, gather honey that comes from a variety of flowers. The amount of honey produced depends on the environmental conditions and the availability of flowers that produce pollen and nectar. The primary honey harvest takes place throughout the fall season. In the research area, 30.1% of beekeepers harvest in October, 74.5% harvest in November, which is the most productive month, and 48.1% harvest in December. The secondary harvest, characterized by lower yields, commences in March and gradually increases from 12% to 27.9% in April. However, there is a notable decline in activity in May, with a production rate of 11.5%.

The seasonality of honey production is marked at two different times of the year in most of the country, in the southeast and coastal regions it is obtained from March to May (spring-summer), generating 40% of production. The second harvest is obtained in the Altiplano and north of the country between September and November (autumn-winter), obtaining the remaining 60% of production.

Honey production in the Yucatán Peninsula occurs during winter and spring from December to June and comes from toothleaf goldeneye (*Viguiera dentata*), tzitzilché (*Gymnopodium floribundum*), and some vines (1, 25). In Michoacán, most beekeeping activity takes place in spring and summer (August and September), with the most significant peak during June. This variation in seasonality by zone around the country allows honey to be available throughout the year.

3.4. Problems in honey production

There is a statistically significant difference ($p = .000$, $\chi 2$) in the complaints of beekeepers. Beekeepers with more than 500 hives express concerns about the lack of available spaces to place their apiaries. In recent years, new beekeepers have emerged who establish their apiaries within a distance of less than 2 kilometers, which invades the flowering spaces. This invasion has resulted in a decrease in production. Meanwhile, beekeepers who have fewer than 500 hives prioritize environmental factors, such as high deforestation and fires, unpredictable and reduced seasons, and the indiscriminate use of pesticides that harm bee populations. They also consider production costs, which have risen due to the expensive sugar, a key ingredient. Additionally, they face challenges related to Africanization, including the need to purchase and exchange queens, as well as the costs associated with treating diseases. Furthermore, they lack support for acquiring extraction equipment and managing hives. The poor condition of roads and tracks also hampers beekeepers from inspecting their hives. Even when support is available, it is often insufficient (Figure 2).

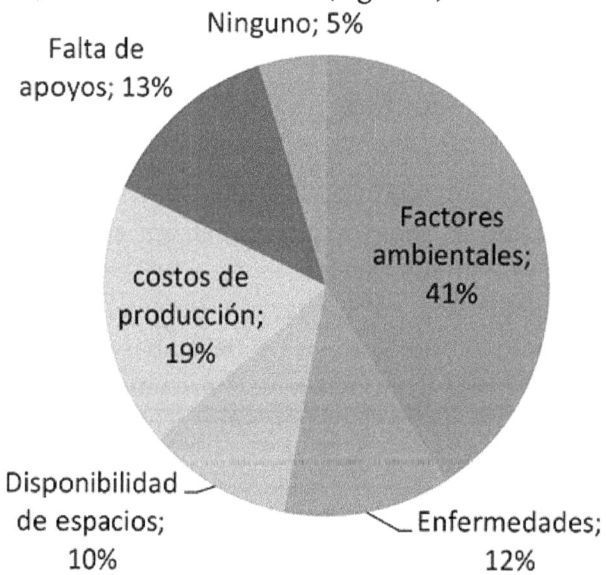

Figure 2. Problems facing beekeepers for honey production (Livestock Research Laboratory, Department of Agriculture Production, CUCSUR, University of Guadalajara).

The primary focus for disease control is on the varroa mite. In Yucatán, beekeepers face a similar predicament characterized by a lack of training and difficult market conditions, which is comparable to the situation observed in the research zone. In Nigeria, beekeepers face various challenges including hive theft, fires, hive abandonment, limited access to advanced technology, lack of technical support, and the aggressive behavior of bees.The number 25 is enclosed in square brackets.

3.5. Bee products

Within the realm of productivity, every single beekeeper (100%) acquires traditional honey. The primary alternative product is beeswax, which is generated by 58.6% of beekeepers. Following in second position are nucleus colonies and propolis, whereas royal jelly, queens, and pollen have a limited impact on the pollination process, if any at all (Table 1). While pollination is not a tangible product, it is a service offered by apiculture. In many regions of the country, it serves as an additional source of revenue. Indeed, in the states of Sinaloa, Chihuahua, and Coahuila, the primary objective of bee exploitation is honey production, with other activities being of secondary importance. In Michoacán, pollination contributes significantly to the economic income of 32.4% of beekeepers. By comparison, the percentage of beekeepers in Michoacán who retrieve beeswax is only 29.1%, whereas in the study area this number is more than twice as high at 58.5%.The number 19 is enclosed in square brackets.

Table 1. Products other than honey obtained by beekeepers in the south and southeast regions of Jalisco (Livestock Research Laboratory, Department of Agriculture Production, CUCSUR, University of Guadalajara).

Alternative product	Frequency	Percentage
Beeswax	107	58.5
Nucleus	52	28.4
Propolis	33	18
Queens	25	13.7
Royal jelly	23	12.6
Pollen	18	9.8

While the production of organic honey results in greater financial gains, it also involves more expenses for equipment, certification processes, and the implementation of specific production protocols to ensure a chemical-free product. Engaging in organic bee farming involves facing certain technical obstacles related to training for the production of high-quality goods and effectively managing resources for the procurement of processing equipment and conducting physicochemical product assessments. These professional practices enhance the competitiveness of the activity. The study zone presents a promising market for beekeepers to explore in the realm of organic honey. The price of organic honey can potentially be 30% more than that of conventional honey, offering a lucrative opportunity for beekeepers.The range of values is from 1 to 19, inclusive. Expanding into other products and engaging in pollination could enhance the income of producers. However, these activities demand more time and since they are not the primary economic focus for over 50% of the beekeepers in the study regions, further diversification is hindered. Consequently, they continue to rely on traditional exploitation, primarily producing honey, beeswax, and bee nucleus colonies, which contradicts the concept of comprehensive exploitation.

3.6. Economic aspects

There is no distinction in how honey is managed, whether it is sold in bottles or buckets, as just 14% of beekeepers sell their products with a label. Nevertheless, a survey conducted in 2012 indicates that in Jalisco, the sales of private label bottled honey account for less than 1% of the total production sold by the producers. This move signifies a significant advancement in reaching the end consumer and enhancing the product's value.The user's text is "[26]".

This type of commercialization has enabled the selling of honey that has been tampered with and even substituting it with high-fructose corn syrup, tricking several consumers who buy it under the impression that it is genuine honey at a significantly reduced cost.[1] The honey sector is considered a commodity due to its limited classification based on color and/or flowering, its sale in large quantities, and the absence of technology that allows for value-added export.The number 27 is enclosed in square brackets.

When analyzing the variance using the Waller-Duncan method, a substantial difference is found in the sale price per kilo of honey. Beekeepers in strata 1, who have up to 25 hives, receive an average of 52.71 pesos. On the other hand, beekeepers in strata 5, who have 501 to 1,000 hives, receive between 37.42 and 40 pesos. An observed tendency indicates that there is a clear correlation between the number of hives and the sales price. This is due to the fact that smaller producers sell their products directly to consumers, resulting in higher prices. In contrast, larger producers sell their products wholesale and typically earn a price of around 40 pesos per kg (Table 2). In 2008, Jalisco was regarded as the highest-earning state, obtaining a price of 30.57 pesos per kilo, above the national average of 24.52 pesos. It is important to highlight that the price increased fourfold from the 1990s to 2008, rising from 5.86 to 24.54 pesos nationwide. This increase might be ascribed to the challenges faced by this sector of production.19

As far as the export of honey, only 6.55% of beekeepers mention exporting honey to Germany. The beekeepers in strata 6 with more than 1,000 hives are the ones who export the most, there being a significant difference ($p < 0.001$ using χ^2) compared with strata 2 with 26 to 50 hives. One of the problems observed in the states in the study is that the production is bought by intermediaries who often pay for the harvest in advance and are responsible for positioning the product on the European market. This is a situation that prevails in countries such as Argentina, where it is reported that more than 95% of honey production is for exportation, and which is handled by only a few actors (three or four companies); the crucial points applied to exportation, such as quality control, storage, transport, and retail outlets form part of the marketing and supply strategies of the exporting companies in the area.[27]

Table 2. Sales price of honey per strata in the south and southeast regions of Jalisco (Livestock Research Laboratory, Department of Agriculture Production, CUCSUR, University of Guadalajara).

Variable		N	Average	Sig.
Sale price of honey in 2011 (kg)	1–25	31	52.71	0.000
	26–50	46	48.00	
	51–100	30	43.50	
	101–500	46	40.26	
	501–1,000	19	37.42	
	1,001 o more	10	40.10	
	Total	182	44.57	

Only around 30% of beekeepers know the quality standards required on the international market. Beekeepers in strata 4, 5, and 6 (more than 100 hives) are better trained in these aspects ($p < 0.66$, χ^2) compared with beekeepers with fewer hives, who also show little interest in the export process considering it to involve too much bureaucracy. Producers need to know the quality standards required by the international market, as well as packaging, packing, and prices be competitive. [28]

3.7. Honey marketing problems

Approximately 40% of the polled beekeepers reported difficulties in marketing their honey. These difficulties include poor prices, with some beekeepers saying that they only break even with their production costs. In addition, street merchants selling counterfeit honey at discounted costs have emerged as unjust competition for beekeepers, since buyers are unaware of the honey's quality and purity. In Argentina, there is a prevalent issue of contaminated honey being sold in the local market. Furthermore, the excessive bureaucratic procedures for exporting and the requirement for middlemen discourage producers from venturing into the foreign market. Beekeepers in stratum 1, 2, and 3 express their apprehension regarding the meager per capita intake of honey, which amounts to a mere 320 grams annually. The number 27 is enclosed in square brackets.

During the production process, 43.7% of the individuals surveyed maintain a record of production expenses. The activity is mostly carried out by beekeepers in stratum 4, 5, and 6, who have more than 100 hives. This is supported by statistical evidence with a significance level of $p < .028$, $\chi 2$. Similarly, Torres[22] noted in Chile that 50% of the individuals interviewed admitted to not keeping written records or sales logs.

Of those beekeepers who do maintain records, not all were able to provide complete information, hence only 38.25% (70) of those interviewed were considered in the calculation of production costs, which include containers, treatments, gas, electricity, equipment repair and maintenance, vehicle maintenance, queen bees, beeswax, labor, feed (sugar, and others), protection equipment, and hive management equipment. In this area, there were significant differences ($p < 0.43$, Waller-Duncan) between the strata, observing that beekeepers with more than 500 hives (strata 5 and 6) have lower production costs (16.43 and 19.62 pesos, respectively), while beekeepers with less than 50 hives (strata 1 and 2) have higher production costs at 46.87 and 34.47 pesos, respectively. Lower production costs in strata with more hives may be directly related to the high volumes of inputs purchased to carry out the beekeeping activity and to group purchases to obtain better prices by buying wholesale.

The exploitation of bees that are more defensive, swarming, and evasive leads beekeepers to make changes in the way they are managed, such as relocating apiaries to more distant locations, thereby increasing the costs of transportation and labor (each worker manages fewer hives per day than when working with European bees), and also the protective equipment required against more defensive bees (coveralls and gloves),[29] and the purchase of queens, which before Africanization was minimal. In addition to this is the cost of bee feed, which in recent years has become one of the major costs, given the excessive increase in the price per kilo of sugar. It is estimated that production costs in managed colonies have increased around 30% in comparison with European bees[4] and because of treatments, particularly for the varroa mite. In spite of this, 66.6% of beekeepers believe bee farming is profitable. However, profit margins vary widely in a range from 5% to 200% as a consequence of such great differences between beekeepers.

3.8. Social aspects

About 90% of beekeepers are members of a beekeepers association; 52% belong to 4 of the 11 associations registered in the study regions (Table 3).

Table 3. Participation in beekeeper associations in the south and southeast regions of Jalisco (Livestock Research Laboratory, Department of Agriculture Production, CUCSUR, University of Guadalajara).

Local livestock association of beekeepers (municipality)	Frequency	Percentage
Sayula	26	14.2
Apiteca	17	9.3
Asociación de Tamazula	22	12.0
Gómez Farías	9	4.9
Atoyac	4	2.2
Asociación San Gabriel	19	10.4
Asociación de Zapotiltic	23	12.6
Asociación de Tapalpa	6	3.3
Zapotlán	24	13.1
Zacoalco	9	4.9
Atemajac de Brizuela	6	3.3
Does not belong to any association	18	9.8
Total	183	100.0

A strategic challenge in the technology field is for small beekeepers to communicate clearly with research bodies, to generate a greater degree of professionalism and scientific rigor to meet the competitive challenges emerging in the industry; such communication is more feasible with producers who participate in organizational bodies. This is an important factor to push the competitive development of apiculture production units toward higher levels of social engagement for economic and productive purposes. It is also important to carry out coordinated actions to achieve a common goal, through the identification and planning of collective actions, and confront the control exerted by intermediaries, which would allow better prices for bee products and lead to the activity no longer being considered as merely for subsistence.[19,30]

Associated beekeepers in the study zones indicate that the support they have received from the association to which they belong consists of guidance for obtaining technical resources and training. Through the producers' alliance, they have been able to obtain government economic resources for the construction and equipping of extraction rooms based on the safety requirements within the honey production process. Beekeepers have opted to associate in various ways to deal with their lack of resources and knowledge; however, the way in which they have become associated has often been linked to obtaining government support, as in the Yucatán Peninsula, beekeeping organizations and cooperatives facilitate the adoption of technology, equipment acquisition, storage, and sale of better quality honey, and are promoted by public institutions and civil organizations.[24]

Associative schemes, whether for productive or commercial purposes, are a valuable tool for beekeepers to achieve their objectives; however, these alone are no guarantee of success; any tool has advantages and disadvantages and being aware of these and analyzing them will avoid any false expectations.

Among the associated beekeepers, 26% think it is unnecessary to make any changes within the operation of the association; however, others mention that changes are required, such as better organization and integration among the members of the associations, referring to greater responsible participation of the assemblies. They also express the need for more resource management and technical support. Nevertheless, they recognize that they have obtained an important benefit by participating in the organization, namely training, and they believe that the honey they produce is recognized for its excellent quality because of their training in best practices in apiary management.

Regarding participation in programs or institutions for support management, close to 80% of beekeepers mention having obtained support from the Secretariat of Agriculture, Livestock, Rural Development, Fisheries and Food (*Secretaría de Agricultura, Ganadería, Desarrollo Rural, Pesca y Alimentación*, SAGARPA), followed by the Secretariat of Rural Development (*Secretaría de Desarrollo Rural,* SEDER), with 13.1%. In addition, 6.6% have received support from the Livestock Productivity Incentive Program (*Programa de Estímulos a la Productividad Ganadera*, PROGAN), which provided economic resources, support for hive identification and payment for technical assistance and training, as well as direct support of 75 pesos per hive to beekeepers of 10 to 175 hives and 75 pesos per hive from 175 to 1,500 hives. This contrasts to the participation in the Secretariat of Social Development (*Secretaría de desarrollo Social*, SEDESOL) program, with 2.7%, which unlike the others promotes social and micro business development. Of the 26% who have not received support from any institution, the majority have less than 50 hives (Table 4).

Table 4. Institutions that have provided support to beekeepers in the south and southeast regions of Jalisco (Livestock Research Laboratory, Department of Agriculture Production, CUCSUR, University of Guadalajara).

Institution	Frequency (beekeepers)	Percentage
SAGARPA	146	79.8
SEDER	24	13.1
PROGAN	12	6.6
FIRA	6	3.3
SEDESOL	5	2.7
DIF	3	1.6
None	26	14.20

Similar participation in support programs is observed in Michoacán and Yucatán, which beekeepers consider is mainly because of there being no requirement for guarantees.[19,24]

3.9. Technological aspects

Out of the beekeepers who were interviewed, 84% reported having access to training, which includes help from the SEDER through PSP technicians. Out of these individuals, 80% claim to have implemented the knowledge they acquired from congresses and training sessions with technicians. They mostly applied this knowledge in disease prevention, feeding, and honey production, following excellent practices. The low percentage of beekeepers with technical or higher education is intriguing. Similarly, in the Alhué commune in Chile, 70% of individuals express a desire to receive education on beekeeping themes, according to Torres. [22] Torres also states that 80% of beekeepers claim to have participated in training classes, while the remaining beekeepers believe themselves to be self-taught.The user's text is "[31]". Unlike in Santa Catarina, Brazil, where beekeepers often do not apply the knowledge they get from various sources, primarily due to engaging in beekeeping as a supplementary occupation and facing financial constraints.The number 32 is enclosed in square brackets. It is worth noting that 47% of individuals in the research areas participate in beekeeping as their main activity, which may explain why they are more inclined to implement innovative practices.

Among the individuals surveyed, 94.5% engage in pest and disease management throughout the months of January, February, June, and August, as seen in Table 5. Varroa mites are the primary issue impacting 91% of their bee population, while foulbrood affects a smaller proportion at 24%, and chalkbrood affects 16%. In Yucatán, the prevalence of Varroa is comparable to that of chalkbrood, with chalkbrood affecting 44.4% of the population.30 India provides treatment for 86.7% of beehives, mostly targeting varroa mites and moths.The number 33 is enclosed in square brackets. Similarly, in Canada, varroa has been identified as the primary factor leading to the mortality of bee colonies during winter, accounting for 85% of reported cases.The number 34 is enclosed in square brackets. Additionally, there have been frequent reports of bee colony mortality (Apis mellifera L.) in the United States, Europe, and Japan. The Varroa destructor mite, along with some viral combinations, has been linked to recent bee colony disappearances, posing a significant threat to bee health.The number 35 is enclosed in square brackets.

Considering that varroa is the primary pathological issue in the research areas, 45% of beekeepers have prioritized the management of mite infestation in bee colonies primarily through the use of Bayvarol®. Another 28.4% utilize natural treatments, while 16% choose for Apivar®. Merely 7.7% refrained from implementing any form of treatment. There is limited research in Mexico on the negative impact of varroa on honey production. However, in Valle de Bravo in Mexico State, colonies that were treated with an acaricide against V. destructor were found to produce a significantly higher amount of honey compared to untreated colonies. It is important to consider that the effect of varroa on bee productivity can be influenced by environmental conditions and the type of bee. [37]

Table 5. Feeding and disease control by month in the south and southeast regions of Jalisco (Livestock Research Laboratory, Dept. of Agriculture Production, CUCSUR, University of Guadalajara).

MONTHS	Bee feeding		Disease treatment	
	Frequency	% Beekeepers	Frequency	%Beekeepers
January	25	13.7	94	51.4
February	27	14.8	39	31.3
March	21	11.5	25	13.7
April	31	16.9	21	11.5
May	74	40.4	33	18
June	139	76	80	43.7
July	145	79.2	49	26.8
August	139	78	54	29.5
September	89	48.6	25	13.7
October	19	10.4	4	2.2
November	1	0.5	4	2.2
December	6	3.3	18	9.8

3.10. Feeding frequency and feed type

Out of the beekeepers surveyed, 96.7% of them offer maintenance feed to their bees. Among these, 81% supply energy feed primarily in the form of syrup, while only 33% provide protein feed. In contrast, in Brazil, 63.6% of beekeepers provide maintenance feed and just 9% provide protein feed.The number 38 is enclosed in square brackets. The utilization of fructose and confectionery waste is a rare occurrence. Within the region of Yucatán, a majority of 77.8% of beekeepers opt to nourish their bees with a sugar syrup, whereas an only 14.8% choose to feed them with honey. The remaining portion of beekeepers utilize granulated sugar as their preferred feeding method.The number 28 is enclosed in square brackets. In California, the utilization of honey and sugar syrup is delineated, and in Chile, the act of feeding with honey is also observed. However, this method poses a risk to the health of the colonies if the honey is not sourced from safe origins.

The feeding frequency ranges from 7 to 10 days (36.1%), 11 to 15 days (45.4%), and 16 to 30 days (17.5%). There are different types of feeders used for feeding, with the most commonly used being a plastic bag (about 27% usage). This is followed by a 1-liter tub (25.7% usage), the Doolittle feeder (18.6% usage), and less commonly the Miller feeder (14.2% usage) and plastic soda bottle (13.7% usage). Feeding and the frequency at which feeding occurs is a management technique that ensures the development of strong colonies during periods of abundant nectar flow, resulting in increased output levels. Beekeepers report that in recent years, changes in rainfall patterns and a lack of flowering have necessitated extended times of feeding for bees, which is a significant expense in bee production. The practice of migratory beekeeping, which is now insignificant in these places, has the potential to decrease feeding expenses. However, it is important to take into account the costs associated with relocating the beehives and the intense competition for suitable locations to establish apiaries.

Most beekeepers in the research oversee contemporary giant or Langstroth hives. However, despite acknowledging the benefits of technological advancements and seeing increased productivity, beekeepers in Ethiopia primarily rely on traditional hives due to their lower cost and lack of knowledge about modern alternatives.The number 40 is enclosed in square brackets. In the northern region of Ethiopia, the average honey production per colony was discovered to be 33 kg in contemporary hives and 16 kg in traditional hives. By just transitioning to a modern hive, honey production more than quadrupled.The user's text is "[41]".

3.11. Quality control

In terms of implementing a quality system, 66% of individuals engage in certain practices, primarily focused on preserving cleanliness in equipment, during harvest and post-harvest activities, and avoiding the use of harmful fuels. Additionally, a smaller percentage opt to use vegetable oil instead of paint to preserve the hives.

During the honey harvesting process, 82% of the interviewed beekeepers utilize a smoker to eliminate the bees from the racks. This is done either on its own or in conjunction with shaking or brushing. They utilize wood chips and corn cobs as their fuel source. This aligns with the approved methods of deterring bees from honeycombs during the harvesting process, which include the use of air, shaking, brushing, and smoke generated by clean fuels. A small proportion (4.9%) of individuals utilize chemical repellents (such as carbolic acid, propionic anhydride, and benzaldehyde), which are limited due to their long-lasting effects on honey and their classification as carcinogenic substances. Using hydrocarbons and their derivatives (such as diesel or liquid gas) or materials including chemicals, paints, resins, or organic waste like manure as fuels is not recommended.The user's text is "[9]". These are crucial factors to contemplate in the honey production procedure to safeguard and potentially enhance Mexico's advantageous standing in the global market.

Of the honey processing equipment, 60% of beekeepers reported having a dedicated extraction room, while the remaining beekeepers mentioned having a designated area or relying on renting or borrowing a room for the extraction procedure. The majority of beekeepers, specifically 66%, report that the rooms in which they carry out their activity are furnished with a supply of running water. Approximately 70.5% of the beekeepers included in the survey own stainless steel equipment, such as an extractor and settling tank. These items are considered essential for maintaining high standards in honey production. Furthermore, 16.4% of beekeepers possess galvanized metal extractors, while 11% utilize alternative materials, including some that are self-made. Additionally, 64.5% of beekeepers employ drip trays during the honey harvesting process to safeguard the supers from potential field contamination. Interestingly, in Michoacán, 66.7% of beekeepers own an extractor, but only 39.5% possess a settling tank. Unfortunately, the specific materials used for these items are not specified. It is worth noting that despite incorporating technology, these beekeepers have not updated it to meet current demands for safe food products.The number 19 is enclosed in square brackets.

Approximately 63.4% of beekeepers utilize a sieving or straining method to filter the honey after extraction. The remaining beekeepers just allow the honey to settle in a tank for 48 hours before bottling it. Conversely, it was noted that 87% of farmers in Michoacán employ honey filtration.The number 19 is enclosed in square brackets. Technological advancement is a crucial aspect that aims to enhance the efficiency of the production process.The user's text is "[38]". This results in a higher quantity of honey and cost reduction through the enhancement of equipment and instruments inside the beekeeping production system.

4. CONCISE SUMMARY

Beekeeping is predominantly performed by older individuals in the southern and southeastern districts of Jalisco, with limited interest from the younger generation and less participation from women. The activity involves participation from both small and large producers, with more than 1,000 hives. However, the majority of participants are small producers who lack sufficient hives to warrant a full-time commitment. This is evident in a significant decrease in honey output that falls below the national average, primarily due to environmental reasons, elevated production expenses, and health issues, with varroa being the primary obstacle faced by farmers. In order to sustain their position as global exporters, it is imperative to develop strategies that focus on differentiating the quality of honey, as there is currently limited diversification and distinction among bee products. Typically, beekeepers lack knowledge about where their products end up, and only have vague information about buyers who purchase significant quantities, with the products primarily being exported. Similarly, there is a lack of awareness on the precise quality standards required by international export markets. The producers' high level of education has motivated them to actively participate in training and technical assistance forums. They have also embraced technological innovation in hive management, as well as in the harvest and post-harvest processes of honey and its by-products. This includes the adoption of stainless steel equipment in the extraction rooms.

REFERENCES

1. Coordinación General de Ganadería. (2010). Situación actual y perspectiva de la apicultura en México. SAGARPA, Claridades Agropecuarias, 34 p.
2. de AgriculturaSecretaría, Ganadería Desarrollo Rural, Pesca y Alimentación (SAGARPA). (2009). Manual de buenas prácticas pecuarias en la producción de miel. Coordinación general de ganadería, SENASICA. 68 p.
3. Güemes-Ricalde F. J.; Villanueva-G. R. y Eaton G. K. D. (2003b) Honey bee production by the Mayans in the Yucatan Peninsula. Bee World, 84 (4):144–154.
4. Guzmán-Novoa, E.; Correa-Benítez, A.; Espinosa-Montaño, L. G. Guzmán Novoa, G (2011). Colonization, impact and control of Africanized honey bees in Mexico, Vet. Méx., 42 (2) 2011.
5. Satizabal E. M. C.; García N. J. M.; Bernal B. G.; Escobar G. J. A. (1986). Caracterización de la apicultura en el valle del cauca y su futuro desarrollo. Acta Agronómica, 36(1) 98–117.
6. Echazarreta G. C. M. (1999). Caracterización de apicultura en la Península de Yucatán. Memorias del Primer foro de proyectos integrales: Sistema producto miel. SISIERRA, 29–43.
7. Villanueva G. R. y Colli-Ucan W.. (1997). La apicultura en la Península de Yucatán, México y sus perspectivas. Apitec, 6, 25–28.
8. Escalante S R. I.; Catalán H. y Galindo L. (2005). Evolución del producto de sector agropecuario mexicano, 1960–2002: algunas regularidades empíricas. Cuadernos Desarrollo Rural, núm. 54, pp. 87–112.
9. de AgriculturaSecretaría, Ganadería Desarrollo Rural, Pesca y Alimentación (SAGARPA). (2009). Manual de buenas prácticas pecuarias en la producción de miel. Coordinación general de ganadería, SENASICA. 68 p.
10. Escalante S. R. I. y Catalán H. 2008. Situación actual del sector agropecuario en México: perspectivas y retos. Economía Informa, 350:7–25.
11. Güemes-Ricalde F. J. y Pat-Fernández. (2001). Condiciones actuales de la apicultura en el estado de Campeche. SISIERRA. 32 p.

12. FAO (2006). Informe Estatal y nacional de fomento Ganadero de México. FAO_SAGARPA.

13. Garibay S. y Zamora E. (2003). Producción orgánica en Nicaragua. Limitaciones y Potencialidades. EDISA, 109 p.

14. Morales T. C. (2003). La apicultura en México, una visión integral del negocio en los mercados globalizados. Memorias, foro nacional sobre la cadena miel (producción, proceso, comercialización y consumo, 37–40.

15. Ibarra M.; Bacells L. J.; Verdencia A. M.; Eng Y. A.; Pérez M. J.; Vázquez B. T.; Lasa G. R.; Recio Z. A. y Linares S. F. (1988). Metodología de la investigación. Editorial Pueblo y Educación. Cuba

16. Hernández S. R.; Fernández C. C. y Baptista L.P. (2007). Metodología de la investigación. Editorial Mc Graw Hill, 4th ed. Mexico.

17. Malhotra N. K. (2008). Investigación de Mercados (un enfoque práctico). Editorial Prentice Hall. Fifth edition. Mexico.

18. Güemes-Ricalde F.J.; Echazarreta-González C.; Villanueva-G. R.; Pat-Fernández J. M. y Gómez-Álvarez R. (2003a). La Apicultura en la Península de Yucatán. Actividad de subsistencia en un entorno globalizado. RMC, 16, 117–132.

19. Ruiz B.J. (2011). Competitividad y política agropecuaria: el caso de la apicultura con potencial exportador en Michoacán, 2000–2009. Thesis to obtain a Master of Sciences degree in Foreign Trade. Bar institute 194 p.

20. Saner G.; Engindeniz S.; Tolon B.; Cukur F. (2004). The economic analysis of beekeeping enterprise in sustainable development: A case study of Turkey. Apiacta, 38: 342–35.

21. Hernández G. Z. (2009). El sector apícola en Tenerife. Mundo Rural de Tenerife, 5:6–9.

22. Torres B. F. J. 2006. Situación técnica comercial de un grupo de productores apícolas de la comuna de Alhué, (RM). Case study. Thesis to obtain a Bachelor's degree in Agronomy. Faculty of Agricultural Sciences, School of Agriculture, Austral University of Chile, 146 p.

23. Ayala A. M. E (2001). La apicultura en la Península de Yucatán: Un acercamiento desde la ecología humana. Thesis to obtain a Master of Sciences degree in Human Ecology. Department of Human Ecology, Center for Research and Advanced Studies of the National Polytechnic Institute Unit (IPN), Mérida. 196 p.

24. Rosales G. M. y Rubio H. A. (2008). Apicultura y organizaciones de apicultores entre los Mayas de Yucatán. Estudios de Cultura Maya XXV, UNAM, 163–183.

25. Tijani B. A.; Ala A. L.; Maikasuwa M. A. and Ganawa N. (2011). Economic Analysis of Beekeeping in Chibok Local Government Area of Borno State, Nigeria. Nigerian Journal of Basic and Applied Science (2011), 19(2): 285–292.

26. Magaña. M. M. A.; Yolanda Beatriz Moguel O. Y. B.; Sanginés G. J. R.; Leyva M, C. E.; Morales. (2012). Estructura e importancia de la cadena productiva y comercial de la miel en México. Rev. Mex. Cienc. Pecu. 3(1):49–64.

27. Pazzi A. (2009). Sector agropecuario y desarrollo rural. El caso del Valle Bonaerense del Río Colorado (Argentina). Doctoral Thesis, Department of Business Administration Rovira I Virgili University, 475 p.

28. Güemes-Ricalde. F. J.; Bautista. J. Yaá, E. 2006. Perfil del mercado para la miel Natural en la unión Europea. SISIERRA/UQROO. 137 p.

29. Guzman-Novoa E. and Page. R. E. (1994). The impact of Africanized bees on Mexican beekeeping. American Bee Journal, 134:101–106.

30. Ojeda L. R. N. y Martín M. M. M. (2008). Factores que explican el funcionamiento de la asociación de apicultores. Memorias del XII Congreso Internacional de la Academia de Ciencias Administrativas A.C. (ACACIA), Tijuana, B.C., Mexico, p: 1–30.

31. Contreras O. M. P. (2007). Caracterización de explotaciones apícolas entre la IV Y X Región de chile, segunda temporada. Thesis to obtain a Bachelor's degree in Agronomy, Faculty of Agricultural Sciences, School of Agronomy, Austral University of Chile, 121 p.

32. Bender C. M. (2006). Estudo da Competitividade da Cadeia Apícola de Santa Catarina: ênfase na análise da dinâmica competitiva dos segmentos produtor e processador da cadeia Dissertation (Master's degree in Economy). Post graduate program in Economy, Federal University of Santa Catarina. Florianópolis. 132 p.

33. Monga K. and Manocha A. (2011). Adoption and constraints of beekeeping in District Panchkula (Haryana), India. Livestock Research for Rural Development. Volume 23, Article No. 103. Consulted January 20, 2016 at http://www.lrrd.org/lrrd23/5/mong23103.htm.

34. Guzmán-Novoa E.; Eccles L.; Calvete Y.; McGowan J.; Kelly P. G.; Correa-Benítez A. (2010). Varroa destructor is the main culprit for the death and reduced populations of overwintered honey bee (Apis mellifera) colonies in Ontario, Canada. Apidologie 41(4):443–450.

35. Schöning G.; Gisder S.; Geiselhard S.; Kretschmann I.; Bienefeld K.; Hilkerand M.; Genersch E. (2011).Evidence for damage-dependent hygienic behaviour towards Varroa destructor-parasitised brood in the western honey bee, Apis mellifera. Journal Experimental Biological 215 (2):264–271.

36. Arechavaleta-Velasco M. E. y Guzmán-Novoa E. (2000). Producción de miel de colonias de abejas (Apis mellifera L.) tratadas y no tratadas con fluvalinato contra Varroa jacobsoni Oudemans en Valle de Bravo, Estado de México. Veterinaria México, 31(4):381–384.
37. Murilhas A. M. (2002). Varroa destructor infestation impact on Apis mellifera carnica capped worker brood production, bee population and honey storage in a Mediterranean climate. Apidologie, 33:271–281.
38. Feitosa F. D. G.; KhanAhmad Saeed, A. K.: Lúcia Maria Ramos, S. L. M. (2004). Nível Tecnológico e Rentabilidade de Produção de Mel de Abelha (Apis Mellifera) no Ceará. RER, Rio de Janeiro, vol. 42, n° 01, p. 171–188.
39. Mussen E. C.; Laidlaw H. H.; Foote L.; Thorp R. W. Gary N. E.; Watkins, L: H. (1987). Beekeeping in California. Division of Agriculture and Natural Resources University of California, 68 p
40. Yirga G, and Mekonen T. (2010). Participatory Technology and Constraints Assessment to Improve the Livelihood of Beekeepers in Tigray Region, northern Ethiopia, Momona Ethiop. J. Sci., 2: 76–92
41. Mekonen T.; Yirga G.; Hailemichael T. and Amare S. (2011). Prospects of beekeeping in the Northern Ethiopian highlands. Scientific Research and Essays Vol. 6(29), pp. 6039–6043, 30 November, (2011).
42. Echazarrreta, G. C. M; Quezada-Euan J. G.; Medina M. L. A. y Pasteur K. L. (1997). Beekeeping in the Yucatan Peninsula. Development and current status. Bee World, 78 (3), 115–127.

08 EXAMINING THE MANAGEMENT OF STINGLESS BEES IN THE WEST-CENTRAL REGION OF MEXICO: A CASE STUDY ON EXTRACTION AND MELIPONICULTURE

1. INTRODUCTION: AN OVERVIEW

Bees are a vital group of pollinators for terrestrial environments [1, 2]. Approximately 73% of cultivated plant species and over 75% of the global vegetation are pollinated by bees [1, 3, 4]. Nevertheless, despite the significance of these group of creatures, there is undeniable proof of their decreasing population, which poses a serious threat to the pollination services they offer [5], as well as the maintenance of ecosystems and agroecosystems [6, 7]. This has significant ecological and economic ramifications. The main contributing factors to this pollination issue include deforestation and fragmentation of forests, the widespread use of agrochemicals, the presence of bee diseases, the introduction of alien species, and the effects of climate change, among other factors [1, 5, 8–18]. For stingless bees, the main reasons for the decline in their population are the removal of wild nests and changes to their natural habitat [19].

There are almost 20,000 bee species, with Apis mellifera being the most studied due to its ecological and productive qualities, as well as its global distribution. Nevertheless, in tropical regions, the stingless bees (Meliponini) are considered one of the most ecologically and socioculturally important species within the Apidae family [2, 16, 20, 22, 23]. The authors of this chapter have conducted a study project in the west-central region of Mexico with the aim of creating solutions for the preservation and sustainable administration of stingless bees. The purpose of this chapter is to outline the inquiry procedure and the findings in relation to a) the variety, understanding, and approaches of local managers of stingless bees; and b) the social process of adopting integrated meliponiculture technology by the managers in the study area. The chapter is segmented into three distinct sections.

The initial segment examines the variety of stingless bees found in Mexico. The second part examines the administration of stingless bees, while the third and final component showcases a case study on the adoption of integrated meliponiculture technology in the Alto Balsas region of Michoacán, Mexico.

2. STINGLESS BEE DIVERSITY IN MEXICO

Stingless bees are classified under the order Hymenoptera, family Apidae, and tribe Meliponini. Globally, there exist 24 genus, 18 subgenus, and an estimated range of 400 to 500 species [20]. The American continent is regarded as the focal point of variety for this particular group. In the case of Mexico, there have been a total of 16 genera and 46 species observed (Table 1), out of which 26% (12 species) are exclusive to the region [24]. Upon examining the specialist literature on the total number of species documented in Mexico, it was found that 43.5% (20 species) are subject to some sort of human management, as indicated in Table 1.

Table 1. Diversity of meliponini in Mexico and useful species.

Species1	Local name	Management	Region of managed species	Reference (for useful species)
1 **Cephalotrigona oaxacana* (Schwarz, 1948)				
2 *Cephalotrigona zexmeniae* (Ayala, 1999)				
3 *Cephalotrigona*zexmeniae (Cockerell, 1912)	E'hol, Tajbak	(*e*)	Peninsula de Yucatán	[25]
4 **Frieseomelitta nigra* (Cresson, 1878)	Sak-Xic' / Abeja zopilota	(*m,e*)	Península de Yucatán, Sierra de Manantlán Jalisco, Cuenca del Balsas Michoacán	[19, 25, 26]
5 **Geotrigona acapulconis* (Strand, 1919)	Colmena de tierra	(*e*)	Cuenca del Balsas Michoacán	[19]
6 **Lestrimelitta chamelensis* (Ayala, 1999)	Abeja limoncilla	(*e*)	Cuenca del Balsas Michoacán	[19]

Species1	Local name	Management	Region of managed species	Reference (for useful species)
7 *Lestrimelitta niitkib* (Ayala, 1999)	Niitkib, Limón kab Limoncillo		Península de Yucatán, Soconusco Chiapas	[25, 27]
8 *Melipona beecheii* (Bennett, 1831)	Xunaan-Kab / Abeja real/Ajau-chab	(*m*)	Península de Yucatán, Soconusco Chiapas, Tabasco, Veracruz	[19, 25, 27, 28, 29, 30]
9 *Melipona colimana* (Ayala, 1999)	Colmena real	(*e*)	Volcán Colima	[31]
10 **Melipona fasciata* (Latreille, 1811)	Colmena real	(*m,e*)	Cuenca Balsas Michoacán, Sierra Atoyac Guerrero	[19, 32, 33]
11 **Melipona lupitae* (Ayala, 1999)		(*e*)	Cuenca Balsas Michoacán	[31]
12 *Melipona solani* (Cockerell, 1912)	Abeja real roja	(*m*)	Soconusco Chiapas	[27]
13 *Melipona yucatanica* (Camargo, Moure, Roubik, 1988)	Tsets.	(*m*)	Península de Yucatán	[25]
14 **Nannotrigona perilampoides* (Cresson, 1878)	Mehenbol / Doncellita prieta / Mumu/ Abeja trompetera	(*m,e*)	Península de Yucatán, Soconusco Chiapas, Sonora y Sinaloa, Cuenca Balsas Michoacán	[19, 25, 27, 28]

Species1	Local name	Management	Region of managed species	Reference (for useful species)
15 *Oxytrigona mediorufa* (Cockerell, 1913)	Pringadora		Soconusco Chiapas	[27]
16 *Paratrigona guatemalensis* (Schwarz, 1938)				
17 *Paratrigona opaca* (Cockerell, 1917)				
18 **Partamona bilineata* (Say, 1837)	Esculcona/ Mordelona	(*e*)	Cuenca Balsas Michoacán	[19]
19 *Partamona orizabaensis* (Strand, 1919)				
20 *Plebeia cora* (Ayala, 1999)				
21 **Plebeia frontalis* (Friese, 1911)	Us-Kaab/ Yaaxich/ mosquito	(*m,e*)	Península de Yucatán, Tehuacán Puebla	[25]
22 **Plebeia fulvopilosa* (Ayala, 1999)	Abeja sapito	(*m,e*)	Cuenca Balsas Michoacán	[19]
23 *Plebeia jatiformis* (Cockerell, 1912)				
24 *Plebeia llorentei* (Ayala, 1999)				
25 *Plebeia manantlensis* (Ayala, 1999)		(*e*)	Colima, Jalisco	[31]
26 *Plebeia melanica* (Ayala, 1999)				
27 **Plebeia mexica* (Ayala, 1999)				
28 **Plebeia moureana* (Ayala, 1999)				

Species1	Local name	Management	Region of managed species	Reference (for useful species)
29 *Plebeia parkeri* (Ayala, 1999)				
30 *Plebeia pulchra* (Ayala, 1999)				
31 *Scaptotrigona mexicana* (Guérin-Méneville, 1844)	Pisil-nekmej/ Abeja congo	(*m*)	Sierra Norte Puebla, HuastecaPotosina-Veracruz, Soconusco	[27, 30, 34]
32 *Scaptotrigona pectoralis* (Dalla Torre, 1896)	Kantsak/Abeja congoalazana	(*m*)	Chiapas. Península Yucatán, Soconusco Chiapas	[25, 27]
33 **Scaptotrigona hellwegeri* (Friese, 1900)	Abeja Bermeja	(*m,e*)	Cuenca Balsas Michoacán y Guerrero	[19, 32, 33]
34 *Scaura argyrea* (Cockerell, 1912)				
35 *Tetragona mayarum* (Cockerell, 1912)				
36 *Tetragonisca angustula* (Latreille, 1811)	Doncellita/ sayulita	(*m*)	Soconusco Chiapas	[27]
37 *Trigona corvina* (Cockerell, 1913)	KurisKab		Península Yucatán.	[25]
38 **Trigona fulviventris* (Guérin-Méneville, 1844)	MuulKab, Culo de buey		Península Yucatán, Soconusco Chiapas	[25, 27]
39 *Trigona fuscipennis* (Friese, 1900)	Kuris-kab, Tamagaza, Basurera		Península Yucatán, Soconusco Chiapas	[25, 27]

Species1	Local name	Management	Region of managed species	Reference (for useful species)
40 *Trigona nigerrima* (Cresson, 1878)	Tamagaza, Basurera		Soconusco Chiapas	[25]
41 *Trigona silvestriana* (Vachal, 1908)				
42 *Trigonisca azteca* Ayala, 1999				
43 *Trigonisca maya* (Ayala, 1999)	Puup, Chachem			
44 *Trigonisca mixteca* (Ayala, 1999)				
45 **Trigonisca pipioli* (Ayala, 1999)	Puup, Chachem, Cepimilla	(*e*)	Península Yucatán, Cuenca Balsas Michoacán	[19, 25]
46 *Trigonisca schulthessi* (Friese, 1900)				
[1]Based on [58]. *Species reported for Michoacán. (*e*): Extraction. (*m*): Meliponiculture.				

Stingless bees are found in tropical and subtropical climates worldwide, as previously stated. In Mexico, the distribution of these species is primarily found in Neotropical regions [35, 36], closely linked to dry tropical and evergreen forests. However, certain species have also been discovered in mountain ecosystems, mesophillic forests, and temperate mixed pine-oak forests [24]. There is a significant variety of Meliponini bees in two specific regions of west-central Mexico: the Pacific coast and the Balsas River Basin. The Balsas River Basin is particularly important for this group since it has a high number of species that are unique to that area. This information is supported by references 19, 24, and 31.

The bees belonging to the Meliponini tribe possess different anatomical features that set them apart from bees that have a functional stinger. In addition, they exhibit a significant decrease in the branching pattern of the front wings, with spurs that are not divided into two parts, and a row of thick, comb-like hairs along the inner edge of the back leg, known as penicillium [24]. Simultaneously, this collective of bees exhibits a wide range of behavioral patterns, such as cleptobiosis or theft, and varying nesting behaviors, including building nests in tree cavities, underground, or in exposed areas similar to termites [37].

Stingless bees play a crucial role in the ecosystem by pollinating around 30 to 50% of plant species in the lowlands of tropical America [22]. It is estimated that in Mexico, over 80% of crops grown for human consumption rely to varying degrees on these pollinators to ensure efficient production [38].

3. MANAGEMENT STRATEGIES OF STINGLESS BEES

3.1. Traditional management

Stingless bees hold significant importance in numerous locations where different methods have been built to manage and raise these insects, from a sociocultural standpoint. They have impacts on social, economic, and religious elements. This activity has been practiced in tropical America, spanning from México to Brazil, since the Pre-Hispanic era [30, 33, 39, 40, 41]. The indigenous people in Mexico and Latin America still possess and utilize the traditional knowledge and management practices related to stingless bees.

Stingless bees in Mexico have historically been controlled in four specific regions: 1) the Yucatán peninsula, 2) the Gulf coast of Mexico, 3) the Pacific coast between Sinaloa and Jalisco, and 4) the Balsas River Basin in Guerrero and Michoacán [28, 33, 42, 43]. Within each of these domains, there are significant management methods and practices involved, ranging from the extraction of derivative products to the breeding process, including a technique known as "meliponiculture."

Traditionally, Mesoamerican meliponiculture has focused on cultivating stingless bees to obtain valuable products that are important for nutrition and medicine. Honey is primarily utilized as a medicinal supplement and remedy for various conditions including ocular infection, fractures, muscle soreness, sprains, cutaneous wounds, as well as gastrointestinal and respiratory illnesses [19, 44]. In the Balsas region, the pollen, known as "pasacuareta," is taken alone or combined with honey to treat respiratory infections and symptoms of weakness or exhaustion. Another significant commodity is the wax, known as "Campeche wax," which served as a lucrative trading asset in colonial times.

Currently, the ancient methods linked to Mesoamerican meliponiculture are preserved in only a limited number of distinct regions within the Mexican tropics. These customs hold great importance in specific regions of Mexico. They are notably significant in the Yucatán peninsula among the Mayas [45, 46], the Sierra Norte of Puebla among the Nahuas and Totonacos [29], the southern region of Veracruz among the Popolucas [34], and the Itzmo de Tehuantepec among the Zapotecos, Mixes, Zoques, Popolucas, and Nahuas [47].

In addition to the conventional practice of meliponiculture, the extraction of products from stingless bees has been recorded in many different areas of Mexico. Bennett (1964) notes that stingless bees have been observed in regions that are not typically considered part of their usual habitat, such as the Sierra Tarahumara and northern Sinaloa. Previous research has shown the existence of the meliponini species, specifically Nannotrigona perilampoides, in these places [48].

Other significant regions for meliponiculture include the western and southwestern parts of Mexico, specifically from Nayarit to the Balsas River Basin, spanning the states of Guerrero, Michoacán, and Morelos [33, 42, 43]. The primary source of honey extraction in these areas was previously the species Scaptotrigona hellwegeri [49]. In the past, this practice held significant economic and socio-political value as it served as a tribute to the Valle de México, the capital of the Mexica Empire [43]. Hendrichs [42] states that the Balsas region was renowned for its honey production, with organizations organizing expeditions to collect honey and beeswax during the dry season (November-December). The author highlights that the extraction procedures in the Balsas region were a deliberate and specialized trade aimed at locating certain nests. However, this activity led to the loss of the nests. Reyes-González and colleagues [19] have recently reported the extraction of products in the same area, namely in the State of Michoacán.

3.2. Technical advancements in management

Despite periods of neglect and replacement by other production activities, like as apiculture, the practice of meliponiculture continues to exist today. Recently, there has been a renewed interest in this specific activity, leading to the development of initiatives aimed at improving traditional meliponiculture to make it more efficient. The ultimate goal is to conserve meliponini and explore alternative productive enterprises. The resurgence of beekeeping is closely tied to the increasing demand for natural, organic, and homeopathic products such as honey, pollen, propolis, and beeswax. An approach called "integrative meliponiculture" has been created, primarily by academic institutions, to improve procedures and manufacturing methods. This has facilitated the implementation of more effective and environmentally-friendly control measures for these insects in their respective habitats. Instances of this contemporary management approach can be observed in the states of Chiapas, Yucatán, Campeche, Guerrero, and Veracruz [27, 32, 46].

Integrative meliponiculture is a comprehensive approach that considers not only the production aspect, but also the preservation and sustainability of viable colonies of stingless bees. It also involves involving families who are interested in participating in this activity and aims to benefit the community as a whole. The primary objective of integrative meliponiculture is to acquire honey, pollen, and propolis from the same nests, while simultaneously promoting specialized ecosystem services such as greenhouse and agricultural field pollination, as well as overall landscape enhancement. When feasible, incorporating environmental education is a crucial component of this activity. Integrative meliponiculture minimizes the damage of natural nests and promotes their preservation, rescue, and reproduction. This action also serves to restrict the flow of traffic and the introduction of non-indigenous species in the regions where the stingless bee naturally occurs. The fundamental problem now faced in the development of meliponiculture in different locations of the world revolves around these ideas.

4. CASE STUDY

4.1. Study area

The Balsas River Basin makes up 6% of Mexico's continental mass and covers various economically important regions of the west-central Pacific coast and south-central area between 17°00′-20°00′ N and 97°30′ y 103°15′ E. This traverses eight States in the Mexican Republic: Morelos (100%) and portions of the States of Tlaxcala (75%), Puebla (55%), México (36%), Oaxaca (9%), Guerrero (63%), Michoacán (62%), and Jalisco (4%) (Figure 1). For the state of Michoacán, this basin can be divided into three subregions, taking an altitudinal criteria: Alto Balsas, Medio Balsas, and Bajo Balsas (Tepalcatepec). The specific area of study falls inside the sub-region of Alto y Medio Balsas, and includes the municipalities of Charo, Madero, Carácuaro, Nocupétaro, Tacámbaro y Tzitzio (Figure 1).

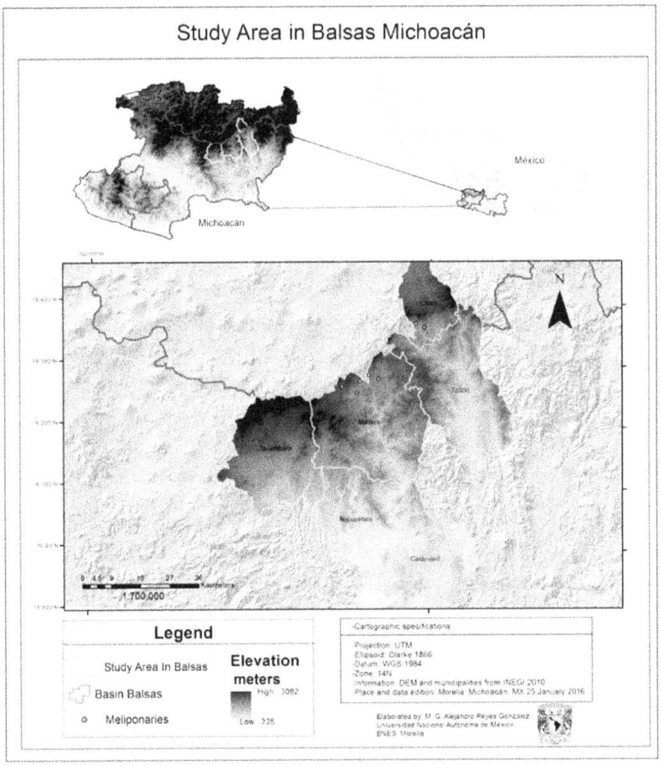

Figure 1. Location map of the study area in the Alto and Medio Balsas, Michoacán, México.

The study area falls in the foothills of the transmexican volcanic belt, flowing from the perennial streams that descend from an altitude of 3000m from the Turicato, Taretio, and El Perdido hills located approximately 10 km to the southeast of Morelia, Michoacán [50]. These subregions of the Balsas in the state of Michoacán has an altitudinal gradient that extends from 3000m to sea level with various climate regions associated with the altitudinal changes. These climate variations principally include temperate, tropical, warm sub-humid, temperate sub-humid, and semiarid [51]. According to Rzedowszki [52], the Balsas basin is considered one of the most biologically diverse regions in the world with a wide range of vegetation types where the principle ecosystems are the tropical dry forest and mixed pine and oak forest [52]. These climatic and vegetative conditions allow for bee species richness where, according to [24], the Balsas zone between Guerrero and Michoacán demonstrates notably high endemism for stingless bees.

Research indicates that temperate zones characterized by a combination of pine and oak forests see significant land use transformation, with forests being converted into agricultural and horticultural areas where avocado cultivation dominates. This situation has significant consequences in the specific area being studied. Similarly, in tropical forest regions, there is a widespread occurrence of land use change for livestock farming, which involves transforming natural vegetation into artificially created grasslands [55]. The perturbations and transformations of natural or mildly transformed landscapes have significant consequences on the presence of stingless bees. These important pollinators experience population decline when their essential sources of pollen, nectar, resins, and nesting niches are eliminated or degraded [56, 57].

4.2. The process of technological appropriation

The process devised to guide sustainable management solutions for the stingless bees in the Alto Balsas region of Michoacán consists of four primary stages. The initial phase involves conducting a comprehensive survey of the many species of stingless bees present in the designated region. The second component is the documenting of the local management techniques of the species found in that area. The third step involves the formation of a working group responsible for selecting species, transferring hives, and doing outreach and communication related to the task. The last stage involves the upkeep and surveillance of the managed beehives.

4.3. Local knowledge and management of the stingless bees in the study area

By conducting comprehensive fieldwork, which included extensive collection in various climatic zones within the study area, engaging in workshops with local apiculturists and "colmeneros" (experts in extracting products from wild nests of stingless bees), and conducting thorough interviews, we were able to document the stingless bee species present in that particular zone, as well as the local knowledge and management practices associated with them. Out of the species mentioned for Mexico, a total of 15 species (which accounts for 32% of the total) were found to be present specifically in the State of Michoacán (as shown in Table 1). It is important to emphasize that there is a large number of different species of stingless bees in the study area. Specifically, 9 species were found, which accounts for 69% of all the species recorded for Michoacán. This information can be seen in Table 2. The altitudinal distribution range of stingless bees spans from 300 to 2000 meters, encompassing the dry tropical forest in the warmest part and the mixed pine-oak woodland in the coldest part of the range. The species confined to the warm zone, namely at altitudes ranging from 300 to 1600 meters, include Scaptotrigona hellwegeri, Trigonisca pipioli, and Frieseomelitta nigra. The species restricted to the temperate zone, namely at altitudes ranging from 1700 to 2000 meters, include Partamona bilineata, Plebeia fulvopilosa, Nannotrigona perilampiodes, and Melipona fasciata. Finally, there are two species, Geotrigona acapulconis and Lestrimelitta chamelensis, that have a wide distribution range of 300-2000 meters.

Table 2. Local knowledge and distribution on stingless bees at the Balsas Region, Michoacan. Based on [19].

Local name	Scientific name	Behavior (local knowledge)	Morphology (local knowledge)	Nesting	Distribution
1) Abeja Bermeja	*Scaptotrigona hellwegeri*	Defensive (gets tangled in the hair and bites).	Intense red dish median bee.	In hollow trunks.	300 –1,600 m Tropical dry forest.
2) Abeja Cepimilla	*Trigonisca pipioli*	Bee type that likes people sweat.	Very small bee.	In hollow trunks, very small nests.	300–1,600 m Tropical dry forest.
3) Abeja Esculcona mordelona	*Partamona bilineata*	Defensive (gets tangled in the hair and bites).	Black middle bee	Aerial and exposed nest as termite mound.	1700–2000 m Oak and pine forest.
4) Abeja Limoncilla	*Lestrimelitta chamelensis*	Docile and attack other bees.	Small dark bee with strong lemon scent.	In hollow trunks.	300–2000 m wide distributions
5) Abeja Sapita	*Plebeia fulvopilosa*	Very docile and timid.	Small dark bee	In hollow trunks and between the trunk and the ground	1700–2000 m Oak and pine forest.

Local name	Scientific name	Behavior (local knowledge)	Morphology (local knowledge)	Nesting	Distribution
6) Abeja Trompetera	*Nannotrigona perilampiodes*	Very docile and timid.	Small bee.	In hollow trunks. The nest entrance is shaped trumpet (made of beeswax).	1700–2000 m Oak and pine forest.
7) Abeja Zopilota	*Frieseomelitta nigra*	Docile	Median dark bee, very bright with white wing tips.	In hollow trunks.	300–1,600 m Tropical dry forest.
8) Colmena real	*Melipona fasciata*	Defensive (gets tangled in the hair and bites).	Similar to *Apis mellifera* in size, color more reddish abdomen that *Apis*.	In hollow trunks	1700–2000 m Oak and pine forest.
9) Colmena de Tierra or Prieta de tierra	*Geotrigona acapulconis*	Very docile and timid.	Medium bee completely dark.	Buried in the ground.	300–2000 m Wide distribution
10) Abeja pintilla	–	–	More small than *Apis mellifera* in size and color similar.	In hollow trunks.	300–1600 m Topical dry forest.
11) Abeja Prieta esculcona	–	–	Black middle bee	Buried in the ground and cavities between the trunk and the ground."	300–1600 m Wide distribution

The research area exhibits a substantial amount of indigenous expertise concerning the native stingless bees, commonly referred to as colmenas or colmenas de palo. It is noteworthy that in addition to the 9 species documented, apiculturists and colmeneros have identified 2 other varieties of stingless bees that have not yet been collected: abeja pintilla and abeja prieta esculcona (Table 2). Consequently, it is possible that the number of different species in the research area is actually more than what has been documented. Table 2 demonstrates that the local apiculturists and colmeneros possess knowledge on the physical traits, nesting behaviors, feeding habits, and defense techniques.

In accordance with the preceding description [19], the local administration in Balsas, located in the State of Michoacán, depends on the direct collection of honey from untamed beehives using basic implements such as axes and machetes. Until the 1980s, the colmeneros depended on specific times of the year to harvest wax and honey, making this activity significant. Beeswax was utilized in the process of candle production and was also available for purchase in different areas as a substance for grafting fruit trees. Honey remains a widely used substance for its therapeutic properties, particularly in treating eye infections, wounds, and bruising. Additionally, it serves as an efficient sweetener. Nevertheless, despite the significance of these bee products, there is a considerable amount of knowledge that has been lost regarding management procedures. This is mostly due to cultural shifts and the decline in bee populations resulting from human-induced changes to the land. Apiculture has been gaining popularity as a widespread and prevalent practice among rural families in the Alto Balsas, Michoacán region. For many years, nearly every household in the region had traditional beehives housing the Castilian honeybee (Apis mellifera). The rustic hives were strategically placed near the keepers' residences or in close proximity to facilitate convenient access for individual household use or local sale. However, the activity of beekeeping in Balsas was significantly affected when the Africanized bee was introduced between 1988 and 1989. As a result, families not only abandoned the use of traditional hives but also stopped keeping European bees. Despite the challenges caused by the introduction of the Africanized bee, the importance of natural bee products and the need to diversify livelihoods in rural areas have led to the implementation of alternative projects. One such project is integrative meliponiculture, which has been developed in the Alto Balsas region of Michoacán.

4.4. Forming the working groups

As previously mentioned, the Balsas region in Michoacán exhibits a notable abundance of stingless bees, however their precise variety and distribution remain incompletely understood. Nevertheless, the investigation undertaken in this study yielded a valuable pilot project concerning the integration of meliponiculture in the municipalities of Nocupétaro and Madero. Under the guidance and influence of the researchers involved in this study, the group "Meliponicultores Michoacanos del Balsas" has transformed into an organization committed to the preservation and sustainable management of stingless bees. It is crucial to highlight that the initiative was driven by genuine interest from the apiculturists, who aim to increase awareness about the significance of these bees and promote their conservation and growth. Additionally, they also recognize the potential for alternative livelihoods that these bees can provide.

4.5. Species selection and hive relocation

We initiated the process by gathering and recording the bees and the precise whereabouts of the hives inside the designated research zone (Figure 1). In the transitional ecotones between temperate forest and dry tropical forest, we found that the most commonly found species were Nannotrigona perilampoides and Plebeia fulvopilosa. Frieseomelitta nigra and Scaptotrigona hellwegeri were more prevalent in the warmer zones. The process of relocating hives was began using these three specific species, in which we transferred the complete hive into specially designed bee boxes. Subsequently, we specifically searched for Melipona fasciata and Scaptotrigona hellwegeri, since these species have been found to be the most suitable for production based on previous experience [32] and the desire of the local management.

At first, we implemented methods and managerial procedures that had been used in other parts of the country for the same type of species found in this area [27, 32]. However, these methods did not produce the desired outcomes, as the rate of successful adaptation and nest retention for Scaptotrigona hellwegeri and Melipona fasciata was only 30%. In addition, species such as Nannotrigona perilampoides and Plebeia fulvopilosa failed to construct their nests. The exercise facilitated a deeper comprehension of the specific biological characteristics involved in the development of each genus we studied.

Through this process, we learned that it was of fundamental importance to use hives at risk of destruction or disappearance either by extraction or habitat destruction, which was causing significant impact on the bee populations. To ensure this means of selection, we developed a series of communication strategies to find out the location of hives and in particular those hives that were at risk.

Through direct dialogue with the inhabitants of the study area and formal presentations in municipal meetings pertaining to rural development organized by the local governing agencies (which were attended by all heads and representatives of the local ejidos, landholders, and communities), we presented the project and activities of the Meliponicultores Michoacanos del Balsas. This allowed us to establish rescue strategies for the nests that were at greatest risk (Figures 2 and 3).

Figure 2. Transfer of wild nest of Plebeia fulvopilosa who was in a oak tree (Quercus sp) at El Herrero, Madero, Michoacán.

Figure 3. Rescue of pillaged nest. One portion of the nest is removed, and recapped for maintenance. Melipona fasciata nest in a pine tree at Pie de la Mesa, Tzitzio, Michoacán.

We also employed a collection technique in which only part of the hive was extracted with only a fragment of the hive resources (honey combs and pollen) that were transported immediately to the new locations where the bee keeping boxes were already prepared for their arrival. This meant very careful extraction from the wild hives to avoid damaging the preexisting internal structure and covering the nest after extraction was completed. If the nest was located in a tree trunk or crevice, the cutting was made with power saw, which allowed for lower disturbance so that the remainder nest was not moved or altered from the original spot and could continue developing in its original location as we have seen in about 80% of nests worked.

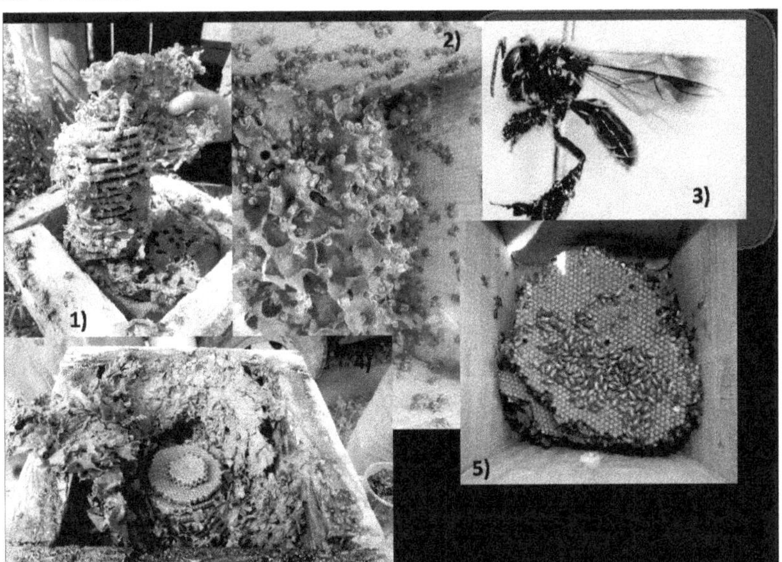

Figure 4. Managed species by the meliponicultores Michoacanos del Balsas: 1) Nannotrigona perilampoides, 2) Melipona fasciata, 3) Frieseomelitta nigra, 4) Plebeia fulvopilosa, 5) Scaptotrigona hellwegeri.

We worked with the following five species of meliponini in the study area: *Frieseomelitta nigra,Melipona fasciata*, *Nannotrigona perilampoides*, *Plebeia fulvopilosa* and *Scaptotrigona hellwegeri* (Figure 4). For each species, we have made every effort to adapt the management strategies and box design to their particular development needs. The smaller boxes have been used with *Frieseomelitta*, *Nannotrigona*, and *Plebeia* genera. By contrast, *Melipona fasciata* requires much larger boxes because the size of their brood combs, pots of storage and bee population (Figure 4).

4.6. Management of beehive upkeep and the process of bringing a product to market

After the hives have settled in the contemporary nesting boxes, it is crucial to regularly inspect them for infestations of parasite flies known as Pseudohyphocera kerteszi. When infestations were found, apple-cider vinegar traps were employed in conjunction with targeted eradication of the flies. In addition, we took measures to guarantee that the nests maintained a consistent supply of resources for the population, such as pollen and honey. If the reserves were depleted, we replaced them with honey from Apis mellifera nests that had ample supplies.

In the hives where there were no external influences or limitations on resources, we collected the following amounts of honey (per hive) in the fall (October-November): Melipona fasciata 1500-2000 ml; Scaptotrigona hellwegeri 800-1000 ml; Nannotrigona perilampoides 100-250 ml; Frieseomelitta nigra 100-250 ml; and Plebeia fulvopilosa 40-80 ml (Figure 5).

Figure 5. Honey harvest of Melipona fasciata by decanting method. Piumo, Michoacán.

In the experience of the working group, though there is a period of flowering in the spring and honey production as a result, products were not extracted from the hives. Instead that was the time when hives were divided to augment the bee and product inventory. As a result of this strategy to allow the hives to take advantage of natural production and leave their nutritional reserves, the monitoring of the nests was not as frequent over the rest of the year when harvesting or dividing were not occurring.

Figure 6. Selling products at fairs and festivals.

The division of the nests was carried out by taking half of the relocated nests (after they were well established) and dividing the storage vessels. In this regard, *Nannotrigona perilampoides* showed the greatest productivity and growth followed by *Plebeia fulvopilosa*. The species that responded less favorably in terms of growth and production after division was *Melipona fasciata* with an estimated 40% success rate after each division.

At the moment, Meliponiculture en el Balsas is being developed on a small scale, it is an innovative activity and has been well received by a group of apiculturists concerned about the rescue and conservation of stingless bees. The products obtained from these bees have a niche in the local and regional market and are sold directly from the producer to the consumer (Figure 5). The honey made into a suspension and is sold as a treatment for ocular infections at a price of around de $50 MN for 25 ml ($3 US dollars). This yields an income of around $120 USD each liter of honey converted into suspension. The locally sourced pure honey is priced at around $1000 MN per liter, equivalent to $65 US dollars. The working group has actively taken part in several commercial events in the regional market, including the State fair and gastronomy events, among others (Image 6). The bee pollen is also available for purchase, blended with honey at a ratio of 100 grams of bee pollen per 1000 milliliters of honey from the Apis mellifera species. This product is advertised as a nutritional supplement that contains a significant amount of protein and has the ability to provide a substantial amount of energy. It is priced at $150 MN (equivalent to $9 US dollars) for every 250 ml of the mixture.

4.7. Challenges and prospects for meliponiculture in Alto Balsas, Michoacán

Throughout this journey, it was imperative to adjust to diverse ecological and biological environments, specifically catering to the distinct requirements of each species we interacted with. The management decisions and resulting treatments were influenced by environmental factors such as flowering periods, seasonality (which leads to limited supplies at certain times of the year), temperature swings, and the presence and abundance of natural nests. These factors indicate the use of adaptive management.

Without a doubt, the most influential factor was the distribution and abundance of wild nests for the species we worked with. For example, *Melipona fasciata* is found exclusively in mountainous zones in areas where mixed pine-oak forest is well conserved. This has been a significant complication for the meliponiculture in the region due to the limited presence of species with the highest production potential. In particular, Melipona fasciata, "la colmena real" is quite scarce and has low resilience to disturbance in the area around nesting sites as it is only found in areas where the vegetation is well conserved and in the hollowed trunks of oaks (Quercus sp).

This species was the sole one that necessitated specific management care. At first, we attempted to utilize the same methods of moving that we did for other species of Meliponini. However, we discovered that these methods were not effective in helping them adjust to their new locations. They either left the sites or were highly vulnerable to being preyed upon by Pseudohyphocera kerteszi. Consequently, the management conducted an experiment by relocating the birds to bigger, rural nest boxes situated in the same areas as their original nests. It was noted that these bees demonstrated adaptability and resilience, allowing them to thrive and reproduce. This led to the conclusion that this species is solitary and tends to choose habitats that are apart from human populations. As a result, there was an exploration of using traditional nests compared to modern ones (hollows in oak trunks) as potential relocation sites for wild nests that are already being managed, whether they are intact or divided. This is similar to the Mayan practice of "jobones," which involves hollowing down oak trunks as described in reference [46].

5. CONCISE SUMMARY

The relationship between human populations and bees has been highly significant across various cultures in America. The association between Mesoamericans and stingless bees has always been quite significant and remains prevalent in certain tropical regions of Mexico. There are multiple management strategies available to exploit the products produced by these insects. Currently, the numbers of stingless bees are decreasing due to environmental deterioration, specifically caused by changes in land use, deforestation, and degradation [19]. Similarly, the conventional wisdom and associated methods of control are also in danger of vanishing due to cultural shifts, economic constraints, and environmental fluctuations.

Given these hazards, it is crucial to enhance the understanding of stingless bees and develop management techniques that align with the specific local circumstances. This study initiative proposes an alternative approach to enhance the breeding and reproduction of these bees in order to support their conservation efforts, while also capitalizing on the benefits they offer. This approach offers a practical way to support the preservation of stingless bees and the valuable ecological benefits they give. It also helps to enhance local expertise and promote sustainable economic activities as an option for rural communities engaged in beekeeping.

REFERENCES

1. Kremen, C. and Tscharntke, T. 2007. Importance of pollinators in changing landscapes for world crops. Proceedings of the Royal Society B: Biological Sciences 274 (1608), 303–313.
2. Brosi, B., Daily, G. and Ehrlich, P. 2007. Bee community shifts with landscape context in a tropical countryside. Ecological Applications 17 (2): 418–430.
3. FAO. 2004. Conservation and management of pollinators for sustainable agriculture, The International Response. In: Freitas B.M., Pereira J.O.P. (eds.). Solitary bees: conservation, rearing and management for pollination. Imprensa Universitária. Fortaleza, Brasil, pp. 2–19.
4. Gallai, N., Salles, J.M., Settele, J. and Vaissiere, B.E. 2009. Economic valuation of the vulnerability of world agriculture confronted with pollinator decline. Ecological Economics 68: 810–821.
5. Taki, H., Kevan, P. G., and Ascher, J. S. 2007. Landscape effects of forest loss in a pollination system. Landscape Ecology 22(10), 1575–1587.
6. Kearns C., Inouye D., and Waser, N. 1998. Endangered Mutualisms: The conservation of plant-pollinator interactions. Annual Review of Ecology and Systematics 29: 83–112.
7. Potts, S. G., Biesmeijer, J. C., Kremen, C., Neumann, P., Schweiger, O., and Kunin, W. E. 2010. Global pollinator declines: trends, impacts and drivers. Trends in ecology & evolution 25(6), 345–353.
8. Allen-Wardell, G., Bernhardt, P., Bitner, R., Burquez, A., Buchmann, S., Cane, J. et al. 1998. The potential consequences of pollinator declines on the conservation of biodiversity and stability of food crop yields. Conservation Biology 12, 8–17.
9. Gonzales, A., J. H. Lawton, F. S. Gilbert, T. M. Blackburn, and I. Evans-Freke. 1998. Metapopulation dynamics, abundance, and distribution in a microecosystem. Science 281:2045–2047.

10. Ricketts, T. H. 2001. The matrix matters: effective isolation in fragmented landscapes. The American Naturalist 158(1), 87–99.

11. Stubbs, C. S., and Drummond, F. A. (eds.). 2001. Bees and crop pollination: crisis, crossroads, conservation. Lanham, MD: Entomological Society of America.

12. Steffan-Dewenter, I., Münzenberg, U., Bürger, C., Thies, C., and Tscharntke, T. 2002. Scale-dependent effects of landscape context on three pollinator guilds. Ecology 83(5), 1421–1432.

13. Steffan-Dewenter, I. 2003. Importance of habitat area and landscape context for species richness of bees and wasps in fragmented orchard meadows. Conservation Biology 17:1036–1044.

14. Kremen, C., Williams, N.M. and Thorp, R.W. 2002. Crop pollination from native bees at risk from agricultural intensification. Proceedings of the National Academy of Sciences 99 (26): 16812–16816.

15. Kremen, C., Williams, N. M., Bugg, R. L., Fay, J. P. and Thorp, R. W. 2004. The area requirements on an ecosystem service: crop pollination by native bee communities in California. Ecology. Letters 7, 1109–1119.

16. Biesmeijer, J. 1997. Stingless bees. Elinkwijk BV Utrecht, Países Bajos, pp 77.

17. Winfree, R., Griswold, T., and Kremen C. 2007 Effect of human disturbance on bee communities in a forested ecosystem. Conservation Biology 21: 213–223.

18. Freitas, B. M., Imperatriz-Fonseca, V. L., Medina, L. M., Kleinert, A. D. M. P., Galetto, L., Nates-Parra, G., and Quezada-Euán, J. J. G. 2009. Diversity, threats and conservation of native bees in the Neotropics. Apidologie 40 (3), 332–346.

19. Reyes-González, A., Camou-Guerrero, Reyes-Salas, O., Argueta, A and A., Casas, A. 2014. Diversity, local knowledge and use of stingless bees (Apidae: Meliponini) in the municipality of Nocupétaro, Michoacan, Mexico. Journal of Ethnobiology and Ethnomedicine 10:47.

20. Michener C.D. 2007. The Bees of the World. Second Edition. The Johns Hopkins University Press, Baltimore EUA, pp 992.

21. Parker, F. D., Batra, S. W. T., and Tepedino, V. J. 1987. New pollinators for our crops. Agricultural Zoology Review 2, 279.

22. Heard, T. 1999. The role of stingless bees in crop pollination. Annual Review of Entomology 44: 183–206.

23. Roubik, D. W., Villanueva, R., Cabrera, E. F. and Colli, W. 1991. Native bees from the Sian Kaáan Biosphere Reserve. In: L. D. Navarro y J. G. Robinson. Biodiversity from the Sian Káan Biosphere Reserve. CIQRO, Chetumal, Quintana Roo, Mexico. pp. 317–320.

24. Ayala, R. 1999 Review of stingless bees from Mexico (Hymenoptera: Apidae: Meliponini). Folia Entomológica Mexicana 106:1–123.
25. González-Acereto, J. A. 2012. The importance of meliponiculture in Mexico, with emphasis on the Yucatan Peninsula. Beekeeping Department, Biological and Agricultural Sciences. UADY.
26. Contreras, E. F. 1999. Knowledge and use of stingless bees (Apidae, Meliponinae) in two communities of the Sierra de Manantlán, Jalisco. Master's Thesis. University of Guadalajara, Mexico.
27. Guzmán-Díaz M. A., Rincón-Rabanales, M. and Vandame, R. 2004. Management and conservation of stingless bees (Apidae: Meliponini). ECOSUR. Tapachula, Mexico pp 40.
28. Bennett, C. F. 1964. Stingless beekeeping in western Mexico. The Geographical Review (54) 85–92.
29. Foster G. M. 1942. Indigenous apiculture among the Popoluca of Veracruz. American Anthropology 44:538–542.
30. Quezada-Euán, J. J., May-Itzá, W. and González-Acereto, J. 2001. Meliponiculture in Mexico: problems and perspective for development. Bee World Num. 84 (4): 160–167.
31. Ayala R, González V, Engel M: Mexican stingless bees (Hymenoptera: Apidae): divesity, distribution, and indigenous knowledge. 2013. In: Vit, P., Pedro-Silvia R. M., and Roubik, D (eds.). Pot-Honey: A Legacy of Stingless Bees. . New York. Springer, pp. 135–152.
32. González-Acereto, J. A. and De Araujo-Freitas, J. Ch. 2005. Manual de Meliponicultura Mexicana. Autonomous University of Yucatan and Fundación Produce Guerrero. Mexico, pp 46.
33. Kent R. 1984. Mesoamerican stingless beeekeeping. Journal of Cultural Geography 4(2): 14–28.
34. Vargas, P. J. P., Vásquez-Dávila, M. A., García-Guerra, T., and Albores, M. L. 2014. Pisilnekmej: a glance at the nahua cosmovision, knowledge and practices on Scaptotrigona mexicana in Cuetzalan, Puebla, Mexico. Etnoecologica 10 (1): 37–40.
35. Halffter,G. 1976. Distribution of insects in the Mexican transition zone. Relations with the North American insect fauna. Folia Entomológica Mexicana 35: 1–64.
36. Yañez-Ordóñez, O., Trujano, M., Llorente, J. 2008. Distribution patterns of the species of the tribe Meliponini (Hymenoptera: Apoidea: Apidae) in Mexico. Interciencia 33:(01) 41–45.

37. Quezada-Euán, J.J. 2005. Biology and use of stingless bees of the Yucatan Peninsula, Mexico (Hymenoptera: Meliponini). Autonomous University of Yucatan. Merida, pp 112.

38. Ashworth, L., Quesada, M., Casas, A. and Oyama, K. 2009. Pollinator-dependent food production in Mexico. Biological Conservation 142: 1050–1057.

39. Crane, E. 1992. The past and the Present Status of Beekeeping with Stingless Bees. Bee World 73 (1): 29–42.

40. Rosso-Londoño J. and Nates-Parra, G. 2005. Meliponiculture: income and environmental services generating activity. LEISA revista de agroecológica 21 (3): 14–16.

41. Cortopassi-Laurino M, Imperatriz-Fonseca VL, Roubik DW, Dollin A, Heard T, Aguilar I, Venturieri G, Eardley C, Nogueira-Neto P. 2006. Global meliponiculture: challenges and opportunities. Apidologie, 37:275–292.

42. Hendrichs PR. 1941. The cultivation of indigenous bees in the State of Guerrero. México Antiguo 5: 365–73.

43. Dixon, C.V. 1987. Beekeeping in Southern Mexico. Conference of Latin Americanist Geographers. Year Book. 13: 68–77.

44. Vit, P., Medina, M.,Enríquez, E. 2004. Quality standards for medicinal uses of Meliponinae honey in Guatemala, Mexico and Venezuela. Bee World 85(1): 25.

45. Villanueva-G R, Roubik DW, Colli-Ucán W. 2005. Extinction of Melipona beecheii and traditional beekeeping in the Yucatán Peninsula. Bee World 86:35–41.

46. González-Acereto, J. A. 2008. Breeding and management of native stingless bees in Mexico. Planeta Impresores. México, pp 177

47. Vásquez-Dávila M. A. 2009. Native Bees of Ethnic Groups of the Isthmus of Tehuantepec, southern Mexico. VI Mesoamerican Congress on Native Bees, Antigua-Guatemala, Guatemala.

48. Cresson E.T. 1878. Descriptions of new species of North American bees. Proceedings of the Academy of Natural Sciences of Philadelphia 30:181–221.

49. Friese H. 1900. Neue Arten der Bienengattungen Melipona Ill, und Trigona Jur. Természetrajzi Füzetek 23:381–394.

50. Toledo, A. 2003. Rivers, coasts, seas. Towards an integrated analysis of hydrologic regions of Mexico. INE, Mexico.

51. Instituto Nacional de Estadística Geografía e Informática (INEGI). 2010. Population and housing census. INEGI, Mexico.

52. Rzedowski, J. 1986. Vegetation of Mexico. Limusa, Mexico, pp 432.

53. Ramirez, L. 2009. Evaluation of land for cultivation of avocado in accordance with local knowledge of landscape in the region of Pico de Tancitaro, Michoacan. Master's Thesis in Geography. Faculty of Philosophy and Literature, UNAM. Mexico, pp 132.

54. Morales Manilla, L.M, Reyes-González, A., Cuevas-García, G. y M. Onchi-Ramuco. 2012. 2011 Inventory of avocado crop in the state of Michoacan. Center for Research in Environmental Geography, UNAM - COFUPRO. Morelia, Mexico, pp 89.

55. Cuevas, G. 2008. Application of a spatial model for developing scenarios of use / land cover in La Huacana, Michoacán. Master's Thesis in Geography. Faculty of Philosophy and Literature, UNAM. Mexico, pp 60.

56. Samejima, H., Marzuki, M., Nagamitsu, T. and Nakasizuka, T. 2004. The effects of human disturbance on a stingless bee community in a tropical rainforest. Biological Conservation 120 (4): 577–587.

57. Lee, Hsiang. S., Navjot, S. and Elmqvist, T. 2001. Bee diversity along a disturbance gradient in tropical lowland forests of south-east Asia. Journal of Applied Ecology 38 (1): 180–192

58. Camargo y S. R. M. Pedro. 2013. Meliponini Lepeletier, 1836. In: Moure, J. S., Urban, D. and Melo, G. A. R., (orgs). Catalogue of Bees (Hymenoptera, Apoidea) in the Neotropical Region - online version. Available at http://www.moure.cria.org.br/catalogue. Accessed Feb/02/2016.

09 BREEDING BUMBLE BEES FOR SCIENTIFIC STUDY AND FINANCIAL GAIN: PRACTICAL AND ETHICAL CONSIDERATIONS

1. INTRODUCTION: AN OVERVIEW

The commercial cultivation of bumble bee colonies commenced in the mid-1980s and has since grown into a global industry valued in the millions of euros [1]. In the 1980s, it was discovered that bumble bees (Bombus spp.) were highly efficient and cost-effective for pollinating greenhouse tomatoes. As a result, they have replaced labor-intensive mechanical pollination methods and hormonal therapies [1]. Currently, the annual production and exportation of colonies has exceeded one million, with countries in Europe, North and South America, and Asia being involved [1]. Besides tomatoes, bumble bees also play a crucial role in pollinating 19 other economically significant crops. The widespread use of bumble bees for pollination has greatly benefited the production of critical food crops by increasing yields, improving fruit quality, and reducing costs [1].

There are about 30 commercial manufacturers globally, but three businesses, namely Koppert Biological Systems and Bunting Brinkman Bees (BBB) based in The Netherlands, and Biobest in Belgium, dominate the market [1]. In order to produce a significant quantity of colonies, it is necessary to establish rearing techniques that can be used on a large-scale industrial level. These techniques entail controlling many parts of the bumble bee life cycle in order to establish colonies at any desired period of the year and in the required quantities. The specific techniques employed by each organization are confidential trade secrets, while the general methodologies have been publicly disclosed [1, 2].

Although the domestication of bumble bees has brought some advantages, it has also unintentionally caused harmful effects on wild populations of bumble bees.

Two of these effects have been well-documented: the release of non-native bees from greenhouses, leading to their establishment in the wild, and the transmission of parasites from these bees to native bee populations.

Nevertheless, I also hold the view that there have been certain pernicious consequences on the manner in which scholarly investigations on bumble bees are being carried out. This study will address my concerns regarding the utilization of commercially generated colonies of Bombus species for essential academic research purposes. It is increasingly prevalent for academics at universities to procure colonies from commercial bumble bee rearing enterprises for their own research and for the research of their graduate students. This is especially true in North America and Europe. The concerns can be classified into two distinct categories: the first two are tangible, while the second pair are intangible.

What are the risks of worsening the problem of parasites and illnesses spreading to wild populations of bumble bees from field research that involve using colonies that were bought?

To what extent do studies conducted on a limited number of species, such as B. terrestris and B. impatiens, accurately reflect the overall population?

Does acquiring and utilizing these colonies imply implicit endorsement of the industry, which could be negatively impacting the indigenous bumble bee populations? This presents a moral dilemma.

Researchers and especially graduate students experiencing a diminished sense of empathy for the creature.

I contend that these are crucial matters that warrant discussion among the entomological and conservation communities. This paper will begin by examining the biology and life cycle of bumble bees. It will next provide a concise overview of the commercial colony rearing sector. Finally, it will assess the likelihood of parasites spreading from infected colonies to bees in both greenhouse and natural environments. Following that, I will examine the utilization of commercially bred colonies for scientific study that aims to achieve pure results, as well as the concerns associated with this approach. In conclusion, I will provide a concise overview of the traditional techniques employed in raising bumble bee colonies and analyze the benefits of utilizing these methods for scientific investigations.

2. BUMBLE BEE BIOLOGY AND LIFE CYCLE

All bumble bees are classified under the genus Bombus, and there are around 250 species of bumble bees found globally [3]. The species primarily inhabit the northern temperate zones, although they can also be found natively in South America. Several species have been imported to New Zealand and Tasmania, where they are considered non-native.

Bumble bees typically exhibit a yearly life cycle. Colonies produce reproductive individuals, including young queens and males, during the late summer. These individuals mate after they have left their original nests [4]. After mating, the newly crowned queens undergo hibernation while retaining the male's sperm in a specialized organ known as the spermatheca. Typically, queens engage in mating just once [5]. In the spring, the queens awaken from hibernation and dedicate their time to gathering nectar and locating an appropriate nesting location. Bumble bees establish their nests in locations where there is already nesting material, such as underground in an abandoned rodent nest, or on the surface in a cluster of dried grass, or near human dwellings, such as in the insulation of buildings [4]. After selecting a nest site, the queen bee proceeds to gather pollen for nourishment to develop her ovaries. Additionally, she collects pollen and places it on the bottom of her nesting hole. She produces cells using wax that is expelled from glands located between the segmented plates on her abdomen. Subsequently, she deposits an initial clutch of 6–10 eggs. In addition, the queen constructs a receptacle known as a honey pot to keep her supply of honey. The initial 3–4 weeks consist of the solitary phase of the life-cycle, during which the queen exclusively engages in foraging and caring for her initial brood.

Bumble bees, like all Hymenoptera, exhibit a haplodiploid genetic system. This means that fertilized eggs give rise to females, whereas unfertilized eggs develop into haploid males. Queens have the ability to manipulate the sex of their progeny by choosing whether to release or retain sperm in their spermatheca. During the initial phase of the season, the eggs produced by the queen will undergo fertilization and subsequently mature into worker bees. After the initial group of workers hatch, the queen stays inside the nest while the workers assume responsibility for activities such as gathering food, protecting the colony, and maintaining its overall condition [4]. During a period of 2-3 months, the colony experiences a rapid growth, and after the worker population reaches a sufficient size to gather enough nutrients, new queens are generated [4]. A prosperous colony has the ability to generate a significant quantity of queens, exceeding 50 in number [6]. The queen, who established the colony, also transitions to depositing unfertilized eggs in order to generate male offspring. Upon the conclusion of summer, the elderly queen, the workers, and the males perish, so leaving behind the fertilized new queens to commence the cycle anew in the subsequent spring.

Bumble bee species, characterized by their diverse tongue-lengths, capability to graze in colder temperatures, and capacity for buzz pollination, are highly efficient pollinators of both wild plants and crops [4].

3. THE COMMERCIAL REARING INDUSTRY

3.1. Economics of the industry

In 2006, Velthuis and van Doorn [1] conducted a comprehensive assessment of the sector and found that the sales of bumble bee colonies had experienced significant development. By 2004, the number of commercial sales had reached approximately one million, starting from its inception in 1988. Since then, it has undeniably grown and the availability of bumble bee colonies is crucial for worldwide tomato production. In 2004, 99,000 acres of greenhouse tomato production were pollinated worldwide by bumble bees, with an estimated value of ~ $15 billion [1]. Exact revenues from bumble bee colony sales are hard to estimate because the companies are private. However, since, for example costs of bumble bee colonies sold by Green Methods.com (https://greenmethods.com/) run from US $109.95–$252.95. The industry as a whole must be worth hundreds of millions of dollars.

3.2. Species used

Traditionally, only species belonging to two subgenera, Bombus sensu stricto and Pyrobombus, have been utilized for commercial breeding purposes. The items are enumerated in Table 1. All of these species are capable of storing pollen. They do so by storing the pollen in wax cylinders located near the brood clumps. This is different from pocket-making species, which store pollen in pockets next to the growing larvae that feed directly on it. Workers of pollen-storing species nourish larvae by regurgitating a blend of pollen and honey. Pollen-storing species can be supplemented with extra pollen, which facilitates their domestication. Out of the species that have been economically raised, two species are primarily utilized: B. terrestris in Europe and B. impatiens in North America (Table 1). Biobest has recently commenced the provision of colonies of B. atratus for utilization in South America. This is a species known as a pocket-maker, characterized by its enormous colonies and its robust and aggressive nature. B. occidentalis ceased commercial production due to a severe infection of Nosema bombi in 1996, most likely originating from wild-caught queens [1].

Table 1. Species and subspecies of bumble bees (*Bombus*) which have been or are currently used for commercial rearing of colonies.

Species	Subspecies	Origin	Used in	Source (Company[1])
Subgenus *Bombus*				
B. terrestris L.				
	B. t. audax (Harris 1780)	Belgium	U.K. only	Biobest

Species	Subspecies	Origin	Used in	Source (Company[1])
	B. t. canariensis Pérez	Canary Islands	Canary Islands	Biobest
	B. t. dalmatinus Dalla Torre	south-eastern Europe, Turkey	Europe	Koppert
	B. t. sassaricus Tournier	Sardinia		Koppert
	B. t. terrestris L.	Europe, Turkey, North Africa, China	Europe, North Africa, Asia, Australasia and Chile	Biobest, Koppert
B. lucorum L.		Europe, Asia	East Asia	
B. ignatius Smith		Belgium	Japan only	Biobest
B. occidentalis Greene[2]		Western North America	Western North America	
Subgenus *Pyrobombus*				
B. impatiens Cresson		Canada, Mexico	North, Central and South America	Biobest, Koppert, Green Methods
Subgenus *Fervidobombus*				
B. atratus Franklin		Argentina	South America	Biobest

[1] Example company.
[2] No longer in production.

4. DANGERS FROM COMMERCIALLY REARED COLONIES

The utilization of commercially raised colonies has become indispensable for the greenhouse tomato business, and any decrease in the pollination services they offer would have significant economic repercussions for the growers. Nevertheless, the extensive utilization of these colonies has brought about certain perils for wild bumble bee populations, which should not be disregarded.

4.1. Escape of non-native species

Bumble bee species have been deliberately introduced to countries where they are not native. As an illustration, four distinct species were brought to New Zealand in 1885 and 1906 with the purpose of facilitating the pollination of red clover [7].

One of these species, B. ruderatus, was subsequently brought to Chile in 1982. Chile already had a native bumble bee species called B. dahlbomi [8]. Nevertheless, there have been occasions where species may have been inadvertently introduced. In 1992, the B. t. audax species, most likely originating from New Zealand, migrated to Tasmania and has since expanded at an average rate of 25 kilometers per year [9]. In 1998, colonies of B. terrestris were brought into Chile from Israel and Belgium with the purpose of using them in greenhouses. Later on, these colonies were utilized for pollinating field crops [8]. There is no question that it is spreading and it is probable that it will become established in the wild. In 2001, I gathered many B. terrestris males from elevated areas located to the south of Santiago. B. terrestris has been brought into Japan since 1992 and has successfully established colonies in the wild [1]. A cause for greater concern is the documented occurrence of hybridization between B. terrestris and B. ignitus in their natural habitat [10]. It is important to mention that the mating of certain subspecies of B. terrestris in captivity happens easily [11]. The eastern species B. impatiens is currently being utilized in unprotected greenhouses located in Alberta and British Columbia inside North America. As an illustration, Ratti and Colla [12] captured a queen and five workers using pan traps placed in fields located at least two kilometers away from the closest greenhouse. Additionally, workers of B. impatiens were observed gathering food on trees adjacent to a commercial greenhouse in Sylvan Lake, Alberta, as reported by Beaudin and Owen (unpublished).

4.2. Common parasitic diseases

The escape and establishment of non-native species in areas poses a genuine threat, as it is impracticable to guarantee that bees will not escape from greenhouses. Furthermore, in regions where greenhouse bees share the same species as the indigenous bees, the escape of these bees can result in the transmission and dissemination of diseases among the local bee populations. Despite the efforts of commercial operations to maintain microparasite-free colonies, a significant number of colonies are likely to be affected [13]. Graystock et al. [13] evaluated the quantities of nine parasites in colonies generated in 2011 and 2012 from the three primary colony producers. They employed molecular techniques to examine the presence of three primary microparasites in bumble bees. These parasites, namely Crithidia bombi, Nosema bombi, and Apicystis bombi, are transmitted by the ingestion of fecal matter and mostly affect adult bees. The researchers also tested for six common parasites found in honeybees: Nosema apis (4), N. ceranae (5), Melissococcus plutonius (6), Paenibacillus larvae (7), deformed wing virus (DWV) (8), and Ascosphaera (9). In addition, they subjected the pollen used to nourish the colonies to pathogen screening. A total of 48 colonies of B. terrestris were analyzed, which were obtained from the three primary suppliers in Europe and imported into the United Kingdom under the assurance of being free from disease. However, upon examination, it was found that 37 out of the 48 colonies (77%) were infected. Among these infected colonies, the presence of parasites ranged from 13% to 56% depending on the specific parasite [13].

Similarly 24 of 25 pollen samples were contaminated with parasites [13]. Also, when bumble bee workers were fed infected pollen or faeces from the commercially produced colonies they would then become infected, and this reduced their survival. The parasites tested were *Crithidia bombi, Apicystis bombi, Nosema bombi*, and *N. ceranae* [13]. It is necessary to test for honeybee diseases as there has been a spread of some of these from honeybees to wild populations of bumble bees [14]. For example Graystock *et al.* [13] found deformed wing virus (DWV) in about 15% of the *B. terrestris* colonies and in 10% of the pollen samples. The recent spread of DWV has been well documented [14, 15]. DWV is endemic in the European honeybee, *Apis mellifera* [15]; however, it is currently remerging as a global epidemic of honeybees. This resulted from the spread of its vector, the mite *Varroa destructor*, from the Asian honeybee *Apis cerana*, which is its normal host [15]. This leap occurred in the middle of the twentieth century and now *V. destructor* is distributed worldwide. The mites are particularly infective because they bear a heavy load of the virus, as it may replicate within the mite or accumulate in the gut; moreover they also inject the virus directly into the hemolymph of the bee [15]. Both the mite and DWV have been implicated as one possible cause of Colony Collapse Disorder in honeybees [15]. Global movement of honeybee colonies brought the Asian and European honeybees into contact and allowed the spread of the host and the virus [15], and the introduction of infected bees into Hawaii, previously free of *Varroa*, led to an increase in virulence of DWV [15]. In sites where infected honeybees and bumble bees coexist, the bumble bees have a higher occurrence of DWV compared to other locations, and they experience reduced rates of survival compared to bees that are not infected [16]. Therefore, it is crucial to treat the potential risk of transmission of infections from domesticated bees with utmost seriousness.

4.3. Spillover of diseases to natural populations: models and data

An implicated aspect in the loss of wild bumble bees is the potential transmission or 'spillover' of infections from greenhouse populations to the wild bees [17–19]. Otterstatter and Thomson [19] have created a model of this. An initial question is, if a single infected colony is introduced into a greenhouse along with other non-infected colonies, how will the infection spread through this closed population? Here we are ignoring the loss of bees to the outside. This can be analysed with a deterministic model of pathogen spread. Here we will consider *C. bombi* which is an intestinal protozoan which spreads both within and between colonies. Once ingested, the parasites attach to wall of the gut using their flagellum. Here they multiply and in a few days, infective cells are shed in the host's faeces. There is no direct transmission from bee to bee, but the infection spreads within a colony when a new host comes into contact with cells on substrates in the nest [20]. In the field, *C. bombi* spreads when bumble bee workers pick up infective cells deposited on flowers by infected bees [17, 19]. The cells are either shed from the body surface of the bee or deposited when the bee defecates [19]. Infection by *C. bombi* can have multiple effects including severely reducing the colony-founding success of queens, the growth rate of established colonies, and worker survival and foraging efficiency [19, and references therein].

4.3.1. Spread in a greenhouse population

A closed population of bees in a greenhouse provides a suitable context for constructing a simple SIR epidemiological model. S, I, and P represent the densities of susceptible bees, infected bees, and infective pathogen particles in the environment, respectively. Let a represent the birth rate of the susceptible population, β represent the transmission rate of pathogen particles, α represent the mortality rate of infected bees, λ represent the rate at which infected bees produce and deposit pathogen particles in the environment, and μ represent the rate at which pathogen particles breakdown in the environment and become non-infective (refer to Table 2). The following assumptions are made: (1) The parameters (a, b, α, λ, μ, γ) remain constant. (2) There is no transmission of the disease within a colony. (3) Each bumble bee can only be infected once. (4) The disease does not transfer directly from one bumble bee to another. Furthermore, the epidemic is expected to last approximately 90 days, specifically from June to August, coinciding with the growth of the colony and affecting exclusively the worker ants.

Table 2.Parameter estimates used by Otterstatter and Thomson (2008) for their model of *Crithidia bombi* spillover to wild bumble bees near greenhouses.

Parameter	Symbol	Value
Birth rate of the susceptible population	a	$0.220\ d^{-1}$
Natural (non-disease) mortality rate	b	$0.183\ d^{-1}$
Disease-induced mortality	α	$0.102\ d^{-1}$
Pathogen production rate	λ	$4.23 \times 10^4\ d^{-1}$
Pathogen decay rate	μ	$12.98\ d^{-1}$
Transmission rate	β	$1.08 \times 10^{-4}\ m^2\ d^{-1}$
Initial host population density	S_0	$0.08\ m^{-1}$
Diffusion coefficient	D	$800\ m^2\ d^{-1}$
From: Otterstatter MC and Thomson JD (2008) Does pathogen spillover from commercially reared bumble bees threaten wild pollinators? *PLoS ONE* 3(7): e2771. doi:10.1371/journal.pone.0002771		

Therefore,

$$\frac{dS}{dt} = (a - b)S + aI - \beta SP$$

$$\frac{dI}{dt} = \beta SP - (\alpha + b)I$$

$$\frac{dP}{dt} = \lambda I - \mu P$$

Thus $S(t)$ = Number (or density) of *susceptible bumblebees* at time t, $I(t)$ = Number (or density) of *infective bumblebees* at time t, and $P(t)$ = Number (or density) of pathogens present in the environment at time t. As an example, if we start with a population of 100 bees with five of these infected, that is, the non-negative initial conditions are $S(0) = S_0 = 100$, $I(0) = I_0 = 5$, and $P(0) \geq 0$. The parameter estimates are those used by Otterstatter and Thomson [19] and are given in Table 2. As shown in Figure 1 the infection sweeps through the population in about 80 days leaving the majority of the bees infected. Clearly in closed populations infections are likely to spread easily.

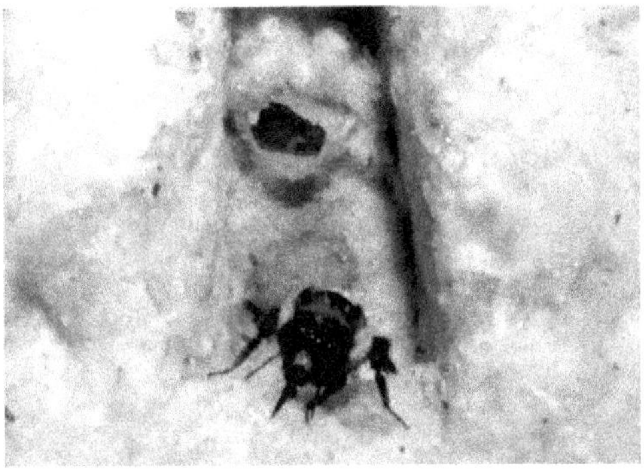

Figure 1. Theoretical course of an infection of Crithidia bombi in a closed greenhouse.

4.3.2. Spatial effects: spillover to wild populations

Horizontal transmission between workers in the field occurs indirectly through the sharing of flowers by infected and vulnerable bumble bees. C. bombi, along with other infections, can be transmitted to wild bee populations when infected bees that were raised in a greenhouse manage to escape and leave short-lived pathogen particles on flowers in close proximity to the greenhouse [19]. Wild bees that are susceptible to infection can acquire the pathogen when foraging near the greenhouse. They become infectious themselves and can bring the pathogen into their own colony and spread it to new flowers.

Otterstatter and Thomson [19] extended the equations provided above to incorporate the propagation of infections not only over time (t), but also across space (x, distance from the initial location).

$$\frac{\delta S}{\delta t} = (a - b)S - \beta SP + D\frac{\delta^2 S}{\delta x^2}$$

$$\frac{\delta I}{\delta t} = \beta SP - (\alpha + b)I + D\frac{\delta^2 I}{\delta x^2}$$

$$\frac{\delta P}{\delta t} = \lambda I - \mu P + D\frac{\delta^2 P}{\delta x^2}$$

Here, the parameters are as defined earlier (Table 2) with the addition of *D*, the dispersal rate of hosts and pathogen particles [19]. Otterstatter and Thomson [19] assumed that wild bees and pathogen particles, which can be picked up and carried on bees' bodies, move about the environment *via* simple diffusion. The model did not include vertical transmission within colonies. Prevalence curves were generated through numerical simulation of the diffusion model using the parameter estimates given in Table 2, to predict the long-term dynamics of *C. bombi* spillover [19]. Initially (*t*= 0–13 wks), pathogen spillover into wild populations is localized around the source; the prevalence of *C. bombi* is about 20% next to greenhouses, and declines to 0% at a distance of roughly 2 km. However in the weeks following, a large wave of infection develops and by 15 weeks, peak prevalence of *C. bombi* near greenhouses has increased to ~ 75%. By 18 weeks, peak prevalence has reached 100%. The wave spreads through the wild bumble bee population at a rate of 2 km/wk [19].

Otterstatter and Thomson [19] validated their model by collecting samples of bumble bee workers from natural populations near greenhouse populations at two locations in southern Ontario, Canada. Based on the parameter estimates utilized, the model exhibited a strong correspondence with the pathogen prevalence observed in the field. The model accurately forecasted the significant decrease in the occurrence of pathogens in the vicinity of greenhouses and closely aligned with the observed prevalence throughout a distance of several kilometers [19].

5. USING BUMBLES BEES FOR RESEARCH

Bumble bees have emerged as significant and extensively utilized organisms for scientific investigation. Their study sheds light on various aspects of biology, encompassing both practical applications like crop pollination and theoretical subjects like the evolution of eusociality and optimal foraging. With the wide accessibility of commercially manufactured colonies, numerous researchers have opted to procure colonies for the purpose of their graduate students' study. The benefits are evident; there is no need for time or effort to raise colonies, and colonies are accessible throughout the year. This is especially advantageous because it allows for research to be conducted during the winter months in areas with moderate climates. Nevertheless, I am of the opinion that there are both possible and existing adverse consequences associated with this strategy.

Commercially manufactured colonies have been utilized for scientific investigation in various domains, including colony formation, worker conduct, foraging conduct, estimation of worker mortality rates, and transplantation experiments to evaluate the pollination effectiveness of distinct subspecies of B. terrestris.

I will now discuss the issues I listed earlier in more detail. The issues fall into two categories, the first two being intangible and the second pair being tangible:

- What is the danger of exacerbating the problem of spread of parasites and pathogens to wild population of bumble bees from field studies using purchased colonies?
- As discussed earlier, the major microparasites of bumble bees are very infectious and liable to spread very rapidly among greenhouse and wild bees. There is a risk of spreading infections from commercially reared colonies to wild populations of bumble bees and this should be avoided.
- How representative are studies done on only a few species, for example *B. terrestris, B. impatiens*?
- There is considerable variation among bumble bee species and subgenera in morphology, behaviour, etc. Concentration on only one or two species for detailed study would seem to be inadvisable.
- Does the purchase and use of these colonies give tacit approval to the industry, which may be having a detrimental effect on the native populations of bumble bees?
- The same scientists may also be decrying the spreading of parasites, etc., and supporting petitions to limit the importation and movement of bumble bee colonies while they are using purchased colonies for their research. Loss of 'a feeling for the organism' [21] by researchers and particularly graduate students.

The term 'a feeling for the organism' was used by the late Nobel-prize winning geneticist Barbara McClintock to express the profound and intuitive comprehension that a biologist might acquire via extensive research on a certain species [21]. The process typically relies on extensive years of observation and research conducted on this particular organism, resulting in valuable insights that cannot be obtained via superficial examination [21]. I am of the opinion that raising bumble bee colonies greatly enhances one's understanding and connection with the creature.

6. THE ART AND SCIENCE OF REARING BUMBLE BEE COLONIES

An advantage of working with bumble bees is the ease and feasibility of comprehensively understanding the creature. An optimal approach to achieve this is to cultivate bumble bee colonies using established techniques that create suitable conditions for queens to commence colonies in captivity in a reasonable and natural manner. This task requires a significant amount of manual labor and must be performed daily throughout the spring and summer seasons, usually starting from mid-April or early May and continuing until mid-August in regions located in the northern latitudes. This entails the regular collection and installation of queens, daily inspection of the nest boxes, provision of food for the queens, and subsequently, the feeding and monitoring of the developing colonies. By conducting the experiment with a substantial number of bees and various species, as indicated in Table 3, one can acquire vital insights into the intricacies of each species and the unique characteristics of each queen.

The approach outlined in this document is derived upon the methodology first established by Plowright and Jay [22], with subsequent modifications made by Owen [23]. This resource is specifically meant to be beneficial for students and researchers who wish to cultivate a significant number (up to 100 or so in a season) of colonies, with minimal effort and in conditions that closely resemble natural settings. The concept is to offer a queen bee with a replication of the natural conditions she would encounter after emerging from hibernation in the spring. This includes a designated nesting place including pre-existing nesting material, as well as a sufficient supply of pollen and nectar.

Table 3. Dates that queens of nine *Bombus* species were collected in the vicinity of Calgary, Alberta in 1985 and 1986, and the number of days until brood one eggs (B1E) lay.

Species	Year	Dates collected	Median	*N*	Days to B1E	Mean	*N*	Success rate
B. nevadensis	1985	May 7–June 7	14 May	23	5 to 17	9.70	12	12/23 = 52%
	1986	May 2–June 14	19 May	23	5 to 24	12.70	11	11/17 - 65%
B. occidentalis	1985	April 30–May 16	08 May	56	3 to 11	6.50	30	30/55 = 54%
	1986	April 19–May 28	12 May	70	3 to 19	7.40	47	47/67 = 70%
B. terricola	1985	May 2–May 16	07 May	27	4 to 12	7.14	14	14/25 = 56%
	1986	May 2–May 28	09 May	30	3 to 13	7.25	16	16/21 = 76%

Species	Year	Dates collected	Median	N	Days to B1E	Mean	N	Success rate
B. californicus	1985	June 5–June 18	11 Jun	5	4 to 11	7.50	2	2/5 = 40%
B. bifarius	1985	May 2–June 7	07 May	24	3 to 10	6.10	15	15/24 = 62%
	1986	May 2–May 28	09 May	12	4 to 12	7.70	7	7/10 = 70%
B. frigidus	1985	May 2–May 16	06 May	42	2 to 8	4.20	21	21/42 = 50%
	1986	April 19–May 28	22 May	30	2 to 13	4.40	16	16/20 = 80%
B. huntii	1985	May 2–June 7	13 May	25	3 to 11	6.80	10	10/13 = 77%
	1986	May 2–May 28	12 May	22	3 to 9	5.80	11	11/18 = 61%
B. perplexus	1985	May 3–May 15	08 May	20	3 to 17	7.10	14	14/19 = 74%
	1986	May 3–May 20	09 May	14	3 to 6	4.80	10	10/14 = 71%
B. ternarius	1985	May 2–June 4	07 May	31	3 to 10	6.50	6	6/27 = 22%
Total				454			242	242/400 = 60%

Fifty four of the queens died after installation (12% mortality) therefore the success rate is calculated for the surviving queens. N, the total number of queens collected for each species; n, the number of queens successfully initiating colonies.

6.1. Queen collection

The optimal approach is to gather recently emerged queens that have just come out of hibernation. The specific date will vary based on the phenology of the species, as indicated in Table 3. These queens are in excellent physical condition and utilizing them results in the most favorable initial rates and more robust growth of the colony. Queen bees that are collecting pollen have already begun building their nests and should not be collected. During the early spring, the queens actively search for food on pussy-willow (Salix spp.) and can be readily located and captured. The bees should be transferred into 5 dram vials equipped with perforated lids to allow for proper ventilation. In order to maintain their temperature, it is necessary to place them on a freezer pack or on bags of ice (covered with a 'J-cloth') in a small cooler. This should be done for a maximum duration of 4-5 hours before they are taken to the laboratory. It is preferable to install the queens as soon as they are brought back to the laboratory. However, if needed, they can be stored in their vials (without any feeding) in the refrigerator at a temperature of around 4°C overnight. Currently, it is possible to measure the weight of bees when they are wet [24], and it is recommended to begin a data sheet for each queen. It is necessary to examine queens for mites. Certain queens may have a significant infestation of mites, totally covering their thorax. These bees should not be used. However, in the case of a small number of mites, they can be manually removed using forceps. The most optimal method for transferring queens is to utilize a vial, and the need to anesthetize them is exceedingly rare, if it exists at all. To safely handle bees, employ broad-tipped forceps and firmly grasp the bee by one of its middle legs.

Table 3 provides the specific dates on which queens of nine Bombus species were gathered in the years 1985 and 1986. During the 1980s, I engaged in extensive collection of bumble bee species in Calgary and the surrounding areas of southern Alberta. Specifically, I focused on two years within this decade. This method allows for the collection of valuable comparison data, including the dates of emergence and the duration till the laying of brood one eggs (B1E). These previous records provided valuable insights on the decreasing abundance of B. occidentalis [25]. The initial or achievement rate under controlled laboratory circumstances for each species is also included in Table 3. Assuming equal establishment of colonies in the wild for all species, it is evident that certain species exhibit superior performance in laboratory settings compared to others. Nevertheless, one can reliably expect the average initial rate to be approximately 60%. If a queen fails to establish a colony within two weeks, it is quite improbable that she would be able to do so. In such cases, she can be preserved using methods like freezing or storing in ethanol for the purpose of genetic investigations. Additionally, her wings can be detached for morphometric analysis [26, 27]. Similarly, monarchs governing colonies can likewise be maintained until their demise.

6.2. Nest boxes and queen installation

The rearing system is characterized by its simplicity, consisting of two wooden boxes: a larger box designated for foraging and defecating (referred to as the 'front' box) and a smaller nesting box (known as the 'back' box). The only crucial aspect of the nesting box is its measurements. If made from half-inch plywood, the outer dimensions should be 4 inches by 4 inches by 2 inches. It is possible to line it with upholsterer's cotton to provide nesting material (see to Figure 2). The boxes should be positioned on a ½" plywood board without being affixed to it, and should be equipped with glass covers. Bees can be housed at a temperature of approximately 20°C and in normal humidity conditions. However, if desired, a moist filter paper or a piece of paper towel can be placed in the nesting box. The illumination conditions or the presence of a light/dark cycle appear to have no significant impact. I have discovered that this method is highly effective in Alberta, where the humidity is typically low. However, in regions with increased ambient humidity during the spring and summer, like as eastern North America, the nesting boxes can be stored in a room with elevated temperature (~30°C) and humidity, similar to the approach used in Chris Plowright's laboratory. Nevertheless, I remain unconvinced of the constant necessity of this, based on my firsthand experience in raising colonies in Alberta. The foraging box is supplied with a honey solution in a 1:1 ratio or a sugar solution in a 60:40 water to sugar volume/volume ratio. Plexiglass bars featuring perforations with a depth of 1 cm are optimal. A pollen mass is placed within the nesting box, measuring approximately 1 cm in diameter and 0.5 cm in height. Pollen aggregates of consistent dimensions can be produced by employing a cork borer and a scalpel. The pollen dough is created by pulverizing fresh and pure pollen (which can be preserved by freezing) together with honey or a solution of sugar, using a mortar. The final paste must possess the optimal consistency, characterized by a balance between stickiness and dryness. Pollen can be acquired from honeybee colonies and can be bought from honeybee breeders. Considering the aforementioned research by Graystock et al. [13], it is imperative to obtain pollen from a source that is free from illness, which may provide challenges in the present day.

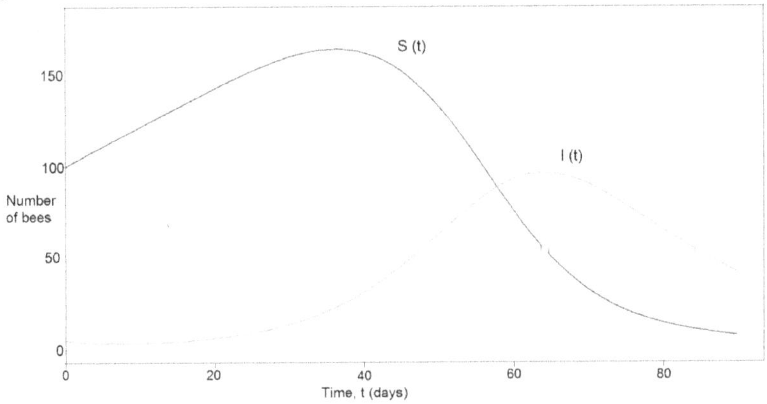

Figure 2. A queen with her first brood clump and a well constructed honey pot. Note also the honey squirted by the queen on the cotton.

6.3. Inspection and feeding

Bees should be examined every other day prior to the commencement of egg-laying. It is advisable to change the plexiglass feeds and clean the unclean bars by washing them completely in hot water. During the days in between, the bar can be easily refilled by using a squeeze bottle filled with a solution of sugar. If there are no eggs present, the pollen mass is substituted; still, it is frequently required to reorganize the upholsterer's cotton as well. If eggs have been deposited or are going to be deposited, little cylindrical masses of the pollen dough are positioned adjacent to the developing brood cluster. Once a queen has initiated egg-laying, she must undergo daily inspection and feeding. This is necessary because she will continue to consume pollen and provide it to the growing larvae. If the queen is inside the rear compartment, tending to her offspring, a gentle tap on it will typically cause the now irritated bee to move into the front compartment. The rear enclosure can then be shifted by a centimeter or two to obstruct her passage back. Alternatively, the glass cover on the front box can be slightly retracted, and then blowing on the entrance hole can be used to coax the queen out. It is recommended to distribute pollen in several little portions throughout the brood area, with the overall quantity being around one quarter of the size of the brood cluster. This is a reliable guideline that is effective and prevents excessive feeding. Any desiccated remnants of pollen are eliminated. If there are any wax pockets present in the brood clump, which is typical for pocket-making species like B. atratus, it is necessary to insert pollen into each pocket. Ensuring this is crucial for pocket-making species, as failure to do so will result in the bees neglecting to nourish their offspring. When dealing with aggressive bees like B. atratus, it is possible to keep the feeding chamber dark and use a red light to feed and handle the bees. This is because they are unable to perceive this particular range of light wavelengths. Curiously, certain species that store pollen will occasionally create pockets within their first-brood larval clumps. Therefore, if these pockets are present, pollen should be given in them.

Several indicators suggest that a queen is preparing to lay eggs or is in the early stages of ovary development. Two significant signals include the presence of chewed pollen lump and the presence of pollen in the feces. The latter is readily visible on the floor of the front box. Furthermore, a 'cavity' frequently develops in the cotton within the rear box, and it is advisable to avoid disturbing it whenever feasible. Occasionally, the pollen mass may be coated with cotton, in which case it is advisable to refrain from disturbing the cavity and instead replace it with a fresh one. Occasionally, wax may accumulate on the pollen mass, and egg cell cups may also be created. It is not necessary to replace the pollen lump in this situation. Instead, just pollen sausages should be applied around the wax. The construction of a honey pot often commences concurrently with or shortly following the queen's oviposition. Certain species, such as B. nevadensis, initiate the construction of their honey pots prior to the commencement of egg-laying. For the queen's survival on consecutive cool and wet days, it is essential to have an ample supply of honey stored. In addition to the honey pot, the queen may occasionally release honey onto the cotton (Figure 2). This is frequently performed either at the time of or shortly before egg deposition. An inherent limitation of upholsterer's cotton is its tendency to get excessively tangled and adhesive. In such cases, it is advisable to promptly eliminate these problematic areas. Ultimately, the queen has the ability to eliminate the pollen mass and produce egg cells directly on the cotton. To facilitate the growth of the brood clump, simply introduce pollen particles near the cells, which will then be assimilated by the queen.

6.4. Deviation from normal development

Occasionally, a queen who is brooding or has already begun the process may exhibit atypical behavior. An especially troublesome behavior is known as front box incubating (FBI), where she incubates on the floor of the foraging box. If detected at an early stage, this condition can be remedied by placing a wire screen (a window screen with a fine mesh) on the floor. Typically, this action will prompt the queen to relocate to the nesting box and lay her initial batch of eggs, or to continue incubating them if they have already been laid. Conversely, the condition has the potential to deteriorate further, resulting in the loss of both the brood and the entire colony. At times, a queen bee will place wax right on the floor of the front compartment and lay her eggs there. It is advisable to encircle the brood clump with cotton and allow it to develop in its current location, rather than attempting to transfer it to the nesting box. Subsequently, a typically unremarkable progression typically ensues.

6.5. Colony transfer

Approximately three weeks after the initial eggs are placed, the first group of worker bees emerge and the honeycomb must be promptly moved to a bigger nesting box to accommodate the further growth of the colony. It is advisable to wait for a few days until the majority, or all, of the initial group of workers have completely enclosed before relocating the comb. Any sufficiently large nesting box will suffice, such as a front box. The comb is positioned centrally, and the remaining area is filled with upholsterer's cotton. This is ideal for colonies intended for outdoor placement and unrestricted foraging, but may not be as practical for colonies intended for laboratory confinement for observation and modification. It is advisable to relocate the colony to a Porous Concrete Hive, often known as a 'perlite' hive, as detailed by Pomeroy and Plowright [28]. An advantage of this particular hive is that it does not require nesting material, making it well-suited for laboratory-based observation and manipulation. A cone of cardboard can be placed on the floor of the hive to line it. This facilitates the ultimate extraction of the comb and simplifies the process of cleansing the hive.

7. CONCISE SUMMARY

The commercial cultivation and utilization of bumble bee colonies is crucial for the cultivation of numerous food crops. Nevertheless, it is important to recognize that there are significant disadvantages associated with their utilization.

REFERENCES

1. Velthuis, H.H.W. and van Doorn, A. (2006) A century of advances in bumblebee domestication and the economic and environmental aspects of its commercialization for pollination. Apidologie 37: 421–451.
2. Hughes, M.J. (1996) Commercial rearing of bumble bees. In: Matheson A. (Ed.), Bumble bees for pleasure and profit. IBRA, Cardiff, pp. 40–47.
3. Williams, P.H., Cameron, S.A., Hynes H.M., Cederberg, B., and Rasmont, P. (2008) A simplified subgeneric classification of the bumblebees (genus Bombus). Apidologie, 39: 46–74.

4. Alford, D.V. (1975). Bumblebees. Davis-Poynter, London.
5. Owen, R.E. and Whidden, T. L. (2013) Monandry and polyandry in three species of North American bumble bees (Hymenoptera: Apidae, Bombus) determined using microsatellite DNA markers. Canadian Journal of Zoology, 91: 523–528.
6. Owen, R.E., Rodd, F.H., and Plowright, R.C. (1980) Sex ratios in bumble bee colonies: Complications due to orphaning? Behavioral Ecology and Sociobiology, 7: 287–291.
7. Gurr, L. (1972) The introduction of bumblebees into North Island, New Zealand. New Zealand Journal of Agricultural Research, 15: 635–638.
8. Ruz, L. (2002) Bee pollinators introduced to Chile: a review. In: Kevan, P. and Imperatriz Fonseca, V.L. (Eds.), Pollinating Bees - The Conservation Link Between Agriculture and Nature, Ministry of Environment, Brasília. pp. 155–167.
9. Hingston, A.B., Marsden-Smedley, J., Driscoll, D.A., Corbett, S., Fenton, J., Anderson, R., Plowman, C., Mowling, F., Jenkin, M., Matsui, K., Bonham, K.J., Ilowski, M., McQuillan, P.B., Yaxley, B., Reid, T., Storey, D., Poole, L., Mallick, S.A., Fitzgerald, N., Kirkpatrick, J.B., Febey, J., Harwood, A.G., Michaels, K.F., Russell, M.J., Black, P.G., Emerson, L., Visoiu, M., Morgan, J., Breen, S., Gates, S., Bantich, M.N., and Desmarchelier, J.M. (2002) Extent of invasion of Tasmanian native vegetation by the exotic bumblebee Bombus terrestris (Apoidea: Apidae). Austral Ecology 27: 162–172.
10. Yoon, H.J., Kim, S.Y., Lee, K.Y., Lee, S.B., Park, I.G., and Kim, I. (2009) Interspecific hybridization of the bumblebees Bombus ignitus and B. terrestris. International Journal of Industrial Entomology, 18: 41–48.
11. Ings, T.C., Raine, N.E., and Chitka., L. (2005A) Mating preference in the commercially imported bumble bee species Bombus terrestris in Britain (Hymenoptera: Apidae). Entomologia Generalis, 28: 233–238.
12. Ratti, C.M. and Colla, S.R. (2010) Discussion of the presence of an eastern bumble bee species (Bombus impatiens Cresson) in western Canada. The Pan-Pacific Entomologist 86: 29–31.
13. Graystock, P., Yates, K., Evison, S.E.F., Darvill, B., Goulson, D., and Hughes, W.O.H. (2013) The Trojan hives: pollinator pathogens, imported and distributed in bumblebee colonies. Journal of Applied Ecology 50: 1207–1215.
14. Villalobos, E.M. (2016) The mite that jumped, the bee that traveled, the disease that followed. Science 351: 554–556.
15. Wilfert, L., Long, G., Leggett, H.C., Schmid-Hempel, P., Butlin, R., Martin, S.J.M., and Boots, M. (2016) Deformed wing virus is a recent global epidemic in honeybees driven by Varroa mites. Science 351: 594–597.

16. Fürst, M.A., McMahon, D.P., Osborne, J.L., Paxton, R.J., and Brown, M.J.F. (2014) Disease associations between honeybees and bumblebees as a threat to wild pollinators. Nature 506: 364–366.

17. Colla, S.R., Otterstatter, M.C., Gegear, R.J., and Thomson, J.D. (2006) Plight of the bumble bee: pathogen spillover from commercial to wild populations. Biological Conservation 129: 461–467.

18. Szabo, N.D., Colla, S.R., Wagner, D.L., Gall, L.F, and Kerr, J.T. (2012) Do pathogen spillover, pesticide use, or habitat loss explain recent North American bumblebee declines? Conservation Letters, 5: 232–239. doi: 10.1111/j.1755-263X.2012.00234.x

19. Otterstatter, M.C. and Thomson, J.D. (2008) Does pathogen spillover from commercially reared bumble bees threaten wild pollinators? PLoS ONE 3(7): e2771. doi:10.1371/journal.pone.0002771

20. Schmid-Hempel, P. (2001) On the evolutionary ecology of host-parasite interactions: addressing the question with regard to bumblebees and their parasites Naturwissenschafen 88: 147–158.

21. Keller, E.F. (1983) A feeling for the organism. W.H. Freeman & Co., New York.

22. Plowright, R.C. and Jay, S.C. (1966) Rearing bumble bee colonies in captivity. Journal of Apicultural Research, 5, 155–165.

23. Owen, R.E. (2001) Rearing bumble bee colonies. Hivelights 14: 15, 22–23.

24. Owen, R.E. (1988) Body size variation and optimal body size of bumble bee queens (Hymenoptera: Apidae). Canadian Entomologist, 120: 19–27.

25. Owen, R.E, Otterstatter, M.C., Cartar, R.V., Farmer, A., Colla, S.R., and O'Toole, N. (2012) Significant expansion of the distribution of the bumble bee Bombus moderatus (Hymenoptera: Apidae) in Alberta over twenty years. Canadian Journal of Zoology, 90: 133–138.

26. Owen, R.E., Whidden, T.L., and Plowright, R.C. (2010) Genetic and morphometric evidence for the conspecific status of the bumble bees Bombus melanopygus and B. edwardsii (Hymenoptera, Apidae). Journal of Insect Science 10: 109.

27. Owen, R.E. and Whidden, T.L. (2013) Discrimination of the bumble bee species Bombus occidentalis Greene and B. terricola Kirby by morphometric, colour and RAPD variation Zootaxa, 3608 (5): 328–344.

28. Pomeroy, N. and Plowright, R.C. (1980) Maintenance of bumble bee colonies in observation hives (Hymenoptera: Apidae). The Canadian Entomologist, 112: 321–326.

INDEX